Sasha Shahbazi
Siavash Javadi

Supporting production system development through Obeya concept

Sasha Shahbazi
Siavash Javadi

Supporting production system development through Obeya concept

An exploratory multiple case study

LAP LAMBERT Academic Publishing

Imprint
Any brand names and product names mentioned in this book are subject to trademark, brand or patent protection and are trademarks or registered trademarks of their respective holders. The use of brand names, product names, common names, trade names, product descriptions etc. even without a particular marking in this work is in no way to be construed to mean that such names may be regarded as unrestricted in respect of trademark and brand protection legislation and could thus be used by anyone.

Cover image: www.ingimage.com

Publisher:
LAP LAMBERT Academic Publishing
is a trademark of
Dodo Books Indian Ocean Ltd. and OmniScriptum S.R.L publishing group

120 High Road, East Finchley, London, N2 9ED, United Kingdom
Str. Armeneasca 28/1, office 1, Chisinau MD-2012, Republic of Moldova, Europe
Managing Directors: Ieva Konstantinova, Victoria Ursu
info@omniscriptum.com

Printed at: see last page
ISBN: 978-3-659-70647-9

Zugl. / Approved by: Eskilstuna, Sweden, 2012

Abstract

The manufacturing industry, as an important part of the European and Swedish economies, faces new challenges with growing global competition. Transforming to a value-based focus will rapidly help overcome these challenges. Investing in tools innovate production system development is crucial to companies adapting their production systems. Those changes can be categorized to incremental and radical ones. This research investigates the Obeya concept as a supporting tool for production system development using incremental and radical approaches. The Obeya concept originated from the Toyota production system and facilitates communication and data visualization through a big meeting space.. Four lean manufacturing companies were studied to find the role of such spaces in production development. Results indicate an opportunity for improving those spaces and their application to change production development projects.

Key words: Production system development, Obeya, Kaikaku, Kaizen, Data visualization

Sasha Shahbazi, Siavash Javadi

MDH, Eskilstuna 2012

Contents

Acknowledgment

We would like to express our gratitude to those who provided us the possibility to complete this thesis. This thesis would never have been completed without their encouragement and devotion.

The authors express sincere appreciation to the case study companies for their extended long-term support, insight and vast patience and knowledge.

We would like to give the special thanks to our supervisor, Professor Mats Jackson from Mälardalen University whose support, guidance and encouragement helped us during our master study and thesis.

We have furthermore to thank Mälardalen University for giving us the opportunity to study and do the research while providing excellent environment and facilities.

Chapter 1: Introduction

This chapter is an introduction to the Obeya concept and discusses where, why and how the concept is required within industry. The chapter also demonstrates research questions, a problem statement and objectives. And finally describes delimitations, thesis outline and area of relevance.

In the last 50 years, Swedish manufacturing companies have struggled to compete with other manufacturing companies around the world due to globalization trends. Consequently, Swedish companies' capabilities in different areas including production, deliveries, logistics and supply chain are key factors in overcoming rivals. On the other hand, companies need to be agile to respond to changes happening, otherwise they would lose the market.

This project investigated Obeya and similar work spaces in Swedish and international industry. The project also utilized the authors' direct experience and observation of implementing an Obeya in Mälardalen University (MDH) and in manufacturing companies as described, compared and analyzed.

Problem statement

By increasing lean production systems adoption trends in new-fashioned industries, especially in Sweden, the requirements to implement the Obeya concept are in place. However, Obeya as a working space integrated to lean production system has not been well conceptualized. On the contrary, meeting places in manufacturing firms play a crucial role with production efficiency and product quality. However, it has not garnered the attention it deserves. As a result, this thesis concentrates on Obeya to quantify the effect they have on production process, product development and competitiveness. Consequently, while there is large demand in product development, Obeya as a lean system product development tool has not been well conceptualized among literature and is not utilized among non-Japanese manufacturing companies.

Objectives and expected result

In regards to background and problems mentioned in the previous section, our thesis objective and expected results are as follows:

The first objective is to gather information and analyze current practice of using Obeya and similar meeting/working spaces in Swedish and international industry.

The second objective is to better understand possible uses of Obeya in change projects with both radical (Kaikaku) and incremental (Kaizen) approaches. Also to investigate opportunities for improving Obeya to support changes in a more effective way.

Research questions

Considering the objectives and expected result, this thesis is broken down to four questions. The first question is intended to provide a background about Obeya to be used as an input for the next questions. The first question is connected to the first objective and the last three questions are designed to accomplish the second objective.

Literary reviews and case studies were completed to answer the first three questions. The final question was answered through results acquired from case studies and observations of authors from those cases.

The research questions are as follows:

1- What is Obeya concept as a part of lean production system?
2- How has Obeya or similar meeting/working places affected daily production tasks including Kaizen tasks?
3- How can the Obeya concept help Kaikaku (radical change) projects?
4- What improvements can be achieved through digitalization of meeting/work places?

Delimitation

Since this thesis is based on a Toyota Production System (TPS) and the lean manufacturing industry, case studies were chosen among manufacturing companies that have applied a lean system. As a result, the focus was on industry leading companies in Sweden that have global manufacturing footprint.

The aim of thesis is neither designing a room/space nor designing software or hardware to be used in this room/space. Focus is on how digitalization and lean tools like visualization can be used to improve the Obeya concept.

Since the study has been done in Sweden, the number of international companies is less than Swedish companies. However, BT, a part of Toyota Material Handling Group, is a leading company in TPS and lean around the world.

Finally, Obeya was developed to help product development process in Japan; however, the real Obeya room does not exist in Sweden according to investigation of authors. Consequently, besides Obeya, daily meeting places were included to increase the thesis validity and case study variety.

This thesis compares the manufacturing industry and academic areas in Sweden, whereas further studies can be performed in other industry and academic areas.

Outline of thesis

CHAPTER 1: Problem statement, objectives and expected result, research questions and delimitations of this thesis.
CHAPTER 2: Describe research methodology and validity
CHAPTER 3: Characterize lean areas in detail according to previous research and literature.
CHAPTER 4: Present case studies and perform analysis.
CHAPTER 5: Suggest results, conclusion and future work.
APPENDIX: Provide related author papers.

Area of relevance and contribution

Several areas is involved in this thesis as follows

- Lean system
  - Visualization tools
  - Continuous improvement (Kaizen)
  - Radical improvement (Kaikaku)
- Project management
  - Product development
  - Production development
  - Competitiveness

Other Obeya areas not included in this thesis can be studied for future development including interior design; develop software or hardware to increase communication, visualization or saving data; innovation and problem solving methods such as brainstorming, knowledge sharing and green systems due to digitalization.

Chapter 2: Research methodology

This chapter describes the authors' scientific views and presents the methodology deployed in this thesis. Moreover, aim, reasons and motivations of using this research approach and data collection is discussed. Finally, the authors discuss the reliability and validity of this approach and process.

Research method

State of the art is reached at the highest level of existing knowledge to make advancements by comprehending requirements, predecessors, techniques, implementation and latest methodology deployment.

The production system development research area is dependent on close communication between academia and industry and the purpose of this thesis is to present results that are valuable for both. Consequently, to achieve the objective, theoretical and practical research was constantly carried out with close collaboration with industry (Fagerström, 2004). Moreover, research results have a direct influence on implementation, long-term learning, recruitment and future development in industries. On the other hand, academic collaboration with industry leads to a significant impact on education, industrial knowledge deployment, student recruitment to academia, long-term learning for both, initiating research projects and quick implementation of research results. In other words, research problems are formulated and solved by academics stance on industrial issues while research results are approved and confirmed by industry.

Since the aim of this research is to add knowledge and provide results for academia and industry, theoretical development was performed employing case studies as the data collection method for this thesis. Case study selection plays a crucial role in research result and consequently, the authors experience helped select them. In this research, each case study followed the same overall purposes as the others (Yin, 1994). As the research is performed at Mälardalen University and the subject is lean manufacturing industries, case studies were chosen representing international leading lean manufacturing companies in Sweden. Expected results includes cross-analysis of companies and analyzing the differences of Obeya's practice in industrial and academic fields. In order to get precise results, close connection between manufacturing industries and academic area was created i.e. practical and theoretical point of views were put together to achieve the best result (Fagerström, 2004).

Research approach

Augmentation to knowledge is the primary purpose of scientific research (Arlbjørn and Halldorsson, 2002) and all research approaches try to promote knowledge.

Generally speaking, research approach imply a conscious scientific reasoning. Furthermore, there are various research approaches within production and logistics, including mathematical modeling, simulation, survey, case studies and interview (Mentzer and Kahn, 1995). Likewise, various data gathering and analysis methods including qualitative and quantitative are performed in production and logistics. Obviously, quantitative techniques swing with numerical data analysis while qualitative techniques work with non-numerical data. Notwithstanding, qualitative and interpretative research is rarely done (Arlbjørn and Halldorsson, 2002, Mentzer and Kahn, 1995, Näslund, 2002)

According to Kova´cs and Spens (2005), research approaches are divided into three main categories and consist of deductive, inductive and abductive. Differences among these three approaches have been indicated in research process, research purpose and time-wise hypothesis and premises. According to Spens and Kova´cs (2006) in order to evaluate the research method, specifications are required including:

- What is the start point of research process, theoretical or empirical study?
- Research aims to develop or assess a theory
- Hypothesis and proposition time-wise introduction
- Research method

Argumentation in deductive approach flows from general law to specific case to result (Andreewsky and Bourcier, 2000, Danermark et al., 2001, Taylor et al., 2002). This research approach is a theory analyzing process to evaluate the existing theories have been developed before. Deductive approach begins with generalization to find out if the theory apply to particular cases. Deductive approach browses the literature in depth, derives hypotheses and propositions, empirically assess them and finally concludes according to the result (Kovács and Spens, 2005, Arlbjørn and Halldorsson, 2002).

On the other hand, inductive flows from empirical case or collected observations to general law and then to results (Andreewsky and Bourcier, 2000, Danermark et al., 2001, Taylor et al., 2002). This approach is a theory development process, which commences with empirical observations or facts on particular cases to generalize a phenomenon. Therefore, it is suitable to develop a new theory (Arlbjørn and Halldorsson, 2002). As a consequence, qualitative research is not inductive, per definition. In an inductive approach, early literature review is not necessary; however; hypotheses are derived from the observation, which later lead to theories (Flint and Mentzer, 2000, Andreewsky and Bourcier, 2000, Kovács and Spens, 2005, Glaser and Strauss, 1967). According to this research approach, empirical study in the earlier stage leads to develop propositions.

Abduction is the systemized creativity or intuition in research to cultivate a new knowledge (Andreewsky and Bourcier, 2000, Taylor et al., 2002). Crucially, creativity makes the difference between previous two approaches and abduction. The abductive approach direction flows from law to result to case (Danermark et al., 2001). Abductive approach helps consider actual phenomena in a new viewpoint by using theories from other fields (Arlbjørn and Halldorsson, 2002, Kovács and Spens, 2005). Additionally, according to Spens and Kova´cs (2006), abductive research may begin with two variant starting points:

1. A "puzzling" and inexplicable observation or an anomaly, respecting to established theory
2. The deliberate application of an alternative theory to explain a phenomenon

Nevertheless, as mentioned before, both cases begin with a real-life phenomenon and observation.

Using theories, empirical observations and data gathering simultaneously in abductive approach, we generate an iterative loop between theory and empirical study that lead to learning. This process is called "theory matching" or "systematic combining" according to Dubois and Gadde (2002) and is used in case study research. Usually, research is started by pre-perception and theoretical experience whereas, in some cases, prior theories do not match with empirical observation. Therefore, a creative iterative process is performed to find a new framework or broaden the earlier observations and theories (Andreewsky and Bourcier, 2000, Taylor et al., 2002, Kovács and Spens, 2005, Dubois and Gadde, 2002).

Ultimately, abductive and inductive approaches aim to develop theories. However, the first objective of abductive approach is to develop a phenomenon in new perspective in the form of hypotheses whereas the purpose of inductive is to get a result from empirical data (Andreewsky and Bourcier, 2000, Kovács and Spens, 2006). Figure 1 demonstrates the three mentioned research approaches and their differences in a very comparable way.

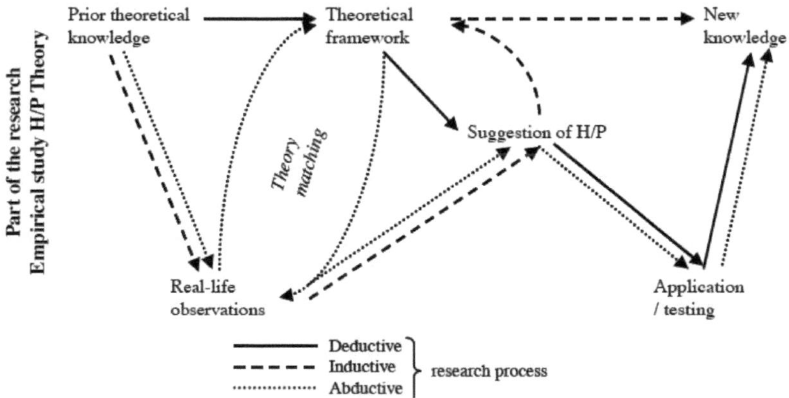

Figure 1 - Different research approaches (Kovács and Spens, 2006)

By incremental trend of qualitative techniques in production and logistics, case studies are becoming more fashionable. A common mistake in this regard is to connect case studies to qualitative techniques, whereas literature indicates quantitative methods can also be linked to case studies (Näslund, 2002, Kovács and Spens, 2006, Ellram, 1996). Many case studies have failed to evaluate their conclusion development due to their failure in the research method phase. As a result, many were labeled as inductive approach while they were deductive. But surprisingly, literature revealed a majority of qualitative studies and several case studies have followed abductive approach in spite of the wrong belief in using inductive approach. However, according to Yin (2003), case studies are study of research not a method (Yin, 2003, Kovács and Spens, 2005). Case studies allow researchers to keep a holistic and considerable characteristic of real events including organizational and managerial process, neighborhood changes, international connections and industrial advancement (Yin, 2003).

At regular intervals, a characteristic of case studies are data collected from various resources and data analyzed simultaneously to present the case in detail in a special subject (Yin, 2003, Ellram, 1993). Nevertheless, case studies do not indicate which research approach has been utilized, but in some case studies, interviews and observations, qualitative research methods are used and can be considered an inductive approach (Flint and Mentzer, 2000, Glaser and Strauss, 1967). Generally, case studies and action research utilize abductive process owing to data gathering and theory development concurrently (Kovács and Spens, 2005, Dubois and Gadde, 2002).

On the other hand, in a different categorization, Arbnor and Bjerke (1994) divided the research approach into three approaches:

The analytical approach is mainly positivistic as demonstrated in Figure 2. Positivistic approach is concerned with reality and uses quantitative techniques including mathematical and statistical methods. In comparison with Kova´cs and Spens results, positivistic approach is a deductive approach and moves from object/low to research/case or reality. Consequently, this approach is mostly utilized in natural science, which based on systematic experiences, tries to empirically test existing theories. Additionally, the positivistic approach is a demonstrative knowledge, which is in contrast with hermeneutics approach (Arbnor and Bjerke, 1994).

In contrast with analytical approach, actors approach is hermeneutic and is concerned with social construction of reality where culture and time are effectual while using interpretive science. In contrast with Kova´cs and Spens results, this approach is inductive, primarily utilized for non-quantitative data and research direction flows from cases to general law (Arbnor and Bjerke, 1994).

The system approach is between actors and analytical while more inclination towards positivistic while the reality is impartially obtainable. This approach utilizes more comprehensive aspect in comparison with the analytical approach and describes a theoretical viewpoint (Arbnor and Bjerke, 1994). It is usually used in complicated frameworks where connected activities ought to be assessed (Salloum, 2010).

In the figure 2 the relationship of mentioned approaches is illustrated.

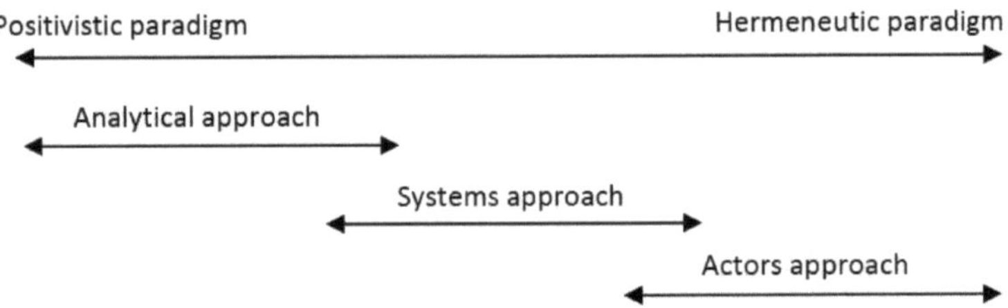

Figure 2- Different research approaches (Arbnor and Bjerke, 1994)

Choosing what approach is appropriate for research is challenging because various elements are involved, including the authors points of view, research question types and problem formulation.

Therefore, according to the mentioned definitions and descriptions about the three different research approaches by Kova´cs and Spens, the abduction research approach is used as this research is followed the abductive principles. This research neither intends to test and assess an existing theory, which is the aim of deductive approach nor deal with quantitative techniques like mathematical and statistical models, which is a basis of analytical approach. Furthermore, this thesis uses empirical cases and observation collection to get results but with an iterative direction between literature and case studies. Thus, this research does not follow inductive or actors approach. The authors do not attempt to characterize hermeneutical or positivistic elements, respecting to the elements' definition.

On the contrary, in this thesis, prior theoretical knowledge in combination with real life observations led to a repetitive loop between theoretical framework and case study. As a consequence, this loop caused generating a new hypothesis and proposition and eventually tests and evaluates each suggestion during the thesis. Moreover, data gathering and theory development was developed simultaneously, which is the typical sign of abductive approach (Kovács and Spens, 2005, Yin, 2003, Arbnor and Bjerke, 1994). On top of all this, the thesis deals with product and production development. Various elements, information and knowledge are used in product and production development including project management, lean production, process development, visualization, etc. Since all of these areas ought to be covered during the research and according to the above-defined approaches, abductive approach is the proper approach in authors' perspective to analyze the subject.

Research process

Research process is defined as the sequential steps a researcher takes in respect to the research approach. Put differently, research process is the indicator of a utilized research approach (Kovács and Spens, 2005). According to the chosen research approach, an iterative process was used. In the process, various theoretical aspects were studied in parallel with empirical observations and data collection. As mentioned previously, this process is "theory matching" or "systematic combining" according to Dubois and Gadde (2002) and is mostly used in case study research and broadens the earlier observation and theories (Kovács and Spens, 2005, Andreewsky and Bourcier, 2000, Taylor et al., 2002, Dubois and Gadde, 2002). As a result, research questions, title and objectives were reviewed during the research conduction. For instance, in the early phase of research, the authors concentrated on the Obeya room as a lean tool. However, during the research's progression, it became clear that an Obeya room, which follows the Toyota principles, is not available or accessible for data gathering in Sweden. Consequently, the authors broadened the research area to include meeting places existing in each company.

Before the thesis started, broad literature research on lean manufacturing had been complete and the authors reviewed available Obeya literature and a general perspective of the subject had been grasped. A number of students participated and gave their ideas and viewpoint about implementing an Obeya room at Mälardalen University in the framework of XPRES lab, an ongoing project to support innovation in Swedish industries. Those discussions led to development ideas to take the first steps. As a consequence, XPRES lab project discussions motivated the authors and gave them a deeper insight into the subject to commence the thesis. A number of lean companies were contacted as potential case studies and four companies were selected as case studies A, B, C and D according to their capabilities and appropriateness to the topic. Respective to the research approach, an iterative process was performed between reviewing literature, visiting the case studies and collecting empirical data. Meanwhile, holding XPRES lab meetings at Mälardalen University with other students expedited the procedure. Further ideas regarding Obeya implementation problems were expressed and discussed, resulting in knowledge sharing between several projects that have something in common with Obeya room. In conclusion, a loop was created consisting of theoretical research, direct case study observation and meeting holding experience at Mälardalen University.

To answer the questions, literature studies and empirical studies were performed. Four companies–A, B, C and D–were visited and interviewed. They were selected according to their commitment to implement and use lean production tools and methods. Sweden was selected as geographical limiting criteria for case selection. Companies A, B, C and D produce material handling equipment, construction material, powertrain systems and parts and trucks' mechanical parts, respectively. Employee numbers at the plants ranged from 160 to 1,800 people. At each company, production processes and meeting places observed were visited and in one case, company C, the authors participated in a morning meeting. All meetings, interviews and visits were documented through voice recording and written documentation.

To answer the first research question, a literature review was conducted. The purpose was to present a complement to the fragmented view of Obeya described in scholarly and non-scholarly literatures. The number of academic articles that has studied Obeya was profoundly few. A number of books and articles about lean production that have very briefly described Obeya. However, in some books such as *The Toyota Way*, the forming Obeya, along with its Reasons for Obeya implementation, have been described. Moreover, due to the uniqueness of the subject, visualization and innovation inside the Obeya room have been studied by only a couple of scholars. Nevertheless, regarding the first three research questions and first objective of the thesis, a literature study was conducted to review the theories to conceptualize the Obeya room, describe the concept more clearly and investigate the Obeya's effect on continuous improvement (Kaizen) and radical changes (Kaikaku). To answer questions two and three, the authors performed empirical study in conjunction with theoretical study. Considering the previously mentioned lack of literature about Obeya rooms, literature about meeting spaces used for production support were also reviewed. The other related areas about Obeya, such as project management, lean production, Kaikaku, Kaizen, visualization, etc. have been studied as well. In the last step, the analysis and comparison between case studies and literature were performed to conclude a valuable result both for academia and industry.

Empirical data collection and case study

Acquiring knowledge about a subject in a direct or indirect observation or experience is called empirical research and used when inadequate data or information is found in literature (Yamamoto, 2010). According to Yin (1994), there are five ways empirical data collection is included

- Experiments: suits for contemporary occurrences in controlled environments when questions stand on "How" and "Why"
- Surveys: focuses on contemporary events when dealing with "Who", "What", How many", "How much" and "Where" questions
- Archival Analysis: appropriate for contemporary and historical cases with "Who", "What", How many", "How much" and "Where" questions
- Histories: perfect for historical cases based on "How" and "Why" questions
- Case study: proper for "How" and "Why" questions in contemporary actions where there is little or no control (Meredith, 1998)

Research question type, investigator's subject control and concentration on contemporary or historical part of the occurrence are the main conclusive factors of choosing a strategy as Yin (1994) stated. The benefit of case study, however, is achievement of an integrated and complete overview of the event by studying it from different perspectives (Gummesson, 2000). According to Meredith (1998), empirical study can build or verify a theory while case studies are used to generate or broaden a theory.

Since this research's aim is to broaden knowledge about the concept and provide a result for academia and industry, theoretical development was performed while employing case studies as the data collection method. Thus, empirically collected data from case studies analyzed in parallel with analysis of the theoretical foundations. The goal of this approach was to ensure practical and theoretical results for academia and industry in the research. Consequently, literary studies results have been written in the second and forth chapters. Project management, lean production, Kaikaku, Kaizen and visualization are included in investigation. Moreover, according to Yin (1994) and Kova´cs and Spens (2005), the start of case studies was of existing knowledge theoretical framework. Therefore, when an apparent perspective of the subject was formed, the research was directed to the case studies. Eventually, the authors entered into the iterative loop between theory and case study as mentioned in the previous sections.

The data collection analysis in this research consists of categorizing and evaluating data and cross-case analysis. Afterwards, the theoretical framework is compared with results achieved from the case studies. In this process, organizing and analyzing the data is essential to the research since without it the researcher would obtain large amounts of information during the research from different sources.

As Yin (1994) stated, researchers have six primary resources of evidence to utilize while none have advantage over the others. As a consequence, they can be used simultaneously and case studies ought to utilize them as much as possible. These resources are documents, archival records, interviews, direct observation, participant observation and physical artifacts.

Interviews are the most substantial and common method of gathering information for case studies, letting the researcher concentrate on the study. However, its disadvantages include selecting wrong or inadequate respondents, choosing poor questions, misinterpretations, incomplete collection and reflexivity as interviewee provides the answers the interviewer wants to hear. Direct observation happens when a researcher visits a place and gathers data. It helps cover the context in real time, although it is time consuming. Occasionally, the observer's existence might lead to change the situation and reality. Participant observation is another observation technique where the researcher takes a part in a specific activity. For example, visiting meeting places in different companies for this research is considered direct observation. However, participating in the meetings is participant observation that has similar pros and cons to direct observation. As a result, the three above-mentioned techniques were used for this research and case studies to gather data, although participant observation was conducted for just one case study due to companies' restrictions.

Interviewing aims to smooth the way theory meets practice (Salloum, 2010). Interviews can be conducted through open-ended, focused, structured and semi-structured forms. In open-ended interviews, the investigator asks for the interviewee's opinion on a subject. This method is used to approve previously information. In the focus method, the interview takes a short time and questions come from a case study protocol. Structured interviews are mostly used in neighborhoods studies where a formal survey is needed.

However, the semi-structured interview was chosen for this research for its flexibility, allowing discussion and allowing for new questions to come up during the interview, which according to Carlsson (1984), increases the validity but may decrease the reliability (Söderberg and Alfredson, 2009). Questions asked during the interviews were based on literary review, objectives and research questions and classified into the four following questions:

- How the meeting space is used?
- For what purposes is it used?
- What are the specifications?
- What kind of improvements can be applied to? What weaknesses have been detected?

Questions considered the purpose of using meeting places, its usage, who participate in meetings, what kind of information is displayed or used in the meeting space, who uses the information in the room, how the information is managed and visualized and finally, the perceived benefits and drawbacks. Furthermore, interviews began with introduction to the research goals. Small parts of the interviews were dedicated to a company presentation and the production process observation. According to the agreement between the authors and interviewees, voices and interviews were recorded as an audio file, were stored on computers and transcribed. Two of the four interviews were in Swedish that were translated to English in a later phase of research.

After visiting all companies, input analysis, information organizing and data reduction were fulfilled to cluster the crucial and related data from the interviews. Afterwards, a cross-case analysis was complete to compare the information and data (Eisenhardt, 1989, Lindlof and Soderberg, 2011). Finally, the theoretical framework was examined in contrast with results achieved from case studies.

To obtain extensive input and different perspectives on the subject, interviewees had different careers and professional backgrounds according to the organizational system of each case. But all had extensive knowledge and experience about the Obeya meeting spaces and their uses.

The interviews were approximately two hours long and since it was a semi-structured interview, various questions comprising a wide range of fields were asked, which led to opportunities to capture different and new subject aspects that had not been planned previously.

Moreover, to ensure the data collection quality and validity, both authors participated in all studies and asked their questions from different perception and perspectives, analyzed data and took notes of important information in the interviews (Lindlof and Soderberg, 2011, Eisenhardt, 1989). To be more precise, as mentioned above, direct observation i.e. participating in morning meetings, documenting gathered data and voice recording were done. Observation by both participants led to different perspectives although may also increase the risk of bias (Yin, 1994).

According to direct observations, different approaches are followed for holding meetings in each case. However, it might be due to different organizational structures and requirements in management. Nevertheless, the general idea and purpose of the meetings were the same.

Reliability and validity

Validity and reliability of research implies the quality and the correctness of the results. In other words, it indicates how research is applicable in the real world and if other scholars can confide in the result (Gummesson, 2000). It implies the same results should be achieved if different researchers investigate the same issue with the same purpose (Yin, 1994).

Likewise, as stated by Yin (1994), the result should be generalizable in either of following ways: statistical generalization or analytical generalization. Statistical generalization employs the number of samples. The more samples there are, the more data that can be collected and as a result, the research is more precise. Analytical generalization deals with the deepness of studies. This describes the depth a researcher focuses into a subject and its features. Hence, few numbers of samples might be covered by quality and in-depth study of issue instead (Yamamoto, 2010).

According to Salloum (2010), the validation of a research stands on five standpoints:

- Internal logic: the research result is formed on previous approved theories exist in the whole work
- Truth: both theoretical and practical aspects of research results should explain real phenomenon
- Acceptance: utilized theories ought to be approved by other scholars
- Application: existence of connection among quality of result and its application
- Innovative thinking: coming up with new approaches or solutions

All aspects are covered in this research. Obeya is approved as a standard development process in Toyota (Liker, 2003). Since Obeya as a working space, integrated to lean production system has not been well conceptualized, it can be assumed as an innovative approach in production development. Meeting places in factories and manufacturing companies play a crucial role with production efficiency and product quality.

In accordance with Westbrook (1994) while the authors were benchmarking the companies, visiting meeting spaces and production processes and interviewing, conversations were recorded as audio files and then transcribed for future use. At each case study, collected information was also frequently written down and notes were organized afterwards to use in reports, analyses and future studies. Moreover, due to documentation and recordings, the information is more reliable because hearing problems no longer exist and hesitations are prevented (Gummesson, 2000). Furthermore, in addition to visiting case studies, other informal information from the interviewing was collected during previous company visits as course projects for cases B and C. Relevant subject information was researched on companies' website and related websites. The authors' technical background, experience and skills have provided significant insight and might provide additional comprehension and interpretation of the cases (Gummesson, 2000, Westbrook, 1995).

The reliability of a research project may not be formulated true or false, but adequate or inadequate. In this regard, the three preliminary-chosen case studies became four to validate the obtained information. However, more companies were contacted but either did not fulfill the research qualifications or could not participate. Companies were chosen from different manufacturing industries, from automotive to construction material in order to cover different aspects and areas. These companies can also be separated according to their size, product, culture and manufacturing processes, which led to maximize learning throughout the research (Stake, 1995, Andersson Schaeffer, 2011). As mentioned in the previous section, to support the research objective, variety of interviewees with high levels of experience and subject knowledge were included. In addition, in all cases the authors observed production lines and processes, they visited meeting spaces and asked questions regarding the meeting places to increase the research validity.

There were at least two-week intervals between each company benchmarking to document and organize the previous data. Meanwhile, the authors tried to address problems or forgotten questions in regards to each case study and prepared themselves for the next case study. As it mentioned in the research approach section, an iterative loop between case study performance and theoretical study produced, which led to a better understanding of the subject, covering mistakes, forgotten questions and difficulties during the research. Consequently, as time went on, the research procedure was developed and negligence was covered for the next company visit. To prove the procedure, Westbrook (1994) states there are several opportunities to correct misunderstandings or wrong assumption during research.

Afterwards, all gains resulting from four case studies and theoretical information were compared in related fields including visualization, project management etc.

At least twice in each month other master's students and a supervisor at the XPRES lab discussed gathered information. These meetings also helped the authors become more confident about their research by receiving feedback from other students and their supervisor. Therefore, it is assumed that this research's reliability and validity is covered from multiple aspects and viewpoints.

Furthermore, a paper was published at an international conference on advancement in production management systems (APMS, 2012). External expert review of the paper confirm and verify the thesis' results.

Chapter 3: Frame of references

This chapter reviews previous literature about integrated subjects in the production system development before addressing the main argument of Obeya in production system development.

Obeya and meeting spaces

The word Obeya in Japanese simply means “big room.” Obeya as a product development support tool was introduced by Uchiyamada, the project chief of Hybrid car development project, in late 90’s (Liker, 2003). This new tool was a part of Toyota achieving a shorter time to market and reduced cost. Obeya is also referred to by authors and scholars with different names such as “war room,” “program room,” “control room” and “the pulse room.” By any name, Obeya is an advanced visual control innovation room (Liker, 2003) that activities and deliverables are outlined and depicted in a visualized format to be discussed in frequent meetings (Lindlof and Soderberg, 2011). However, despite many advantages and applications in industrial product and production development, Obeya has not been covered comprehensively in academia (Lindlof and Soderberg, 2011).

Since Prius development, Obeya has become a standard support tool for vehicle development process at Toyota (Liker, 2003, Andersson Schaeffer, 2011) and a first step toward lean product development (Lindlof and Soderberg, 2011). Oppenheim (2004) states, “One of four success factors and metrics for lean principle is availability of a large, comfortable ‘War Room’ suitable for VSM, for the Program duration.”

In the previous vehicle development projects prior to Obeya, the chief engineer had to visit people’s offices in different departments (virtually or physically) to consult to others about a subject (Liker, 2003). Conversely, in the new system, cross-functional groups of experts from various divisions including production, purchase and product development gather regularly in a big room to review progress and discuss key decisions (Liker, 2003, Andersson and Bellgran, 2009, Söderberg and Alfredson, 2009). Andersson Schaeffer (2011) states that Obeya saves the time as there is no need to move to conference rooms because you have the name and face in front of you to ask questions. In Obeya, it is not solely the chief engineer who controls ongoing affairs and decisions but all people contribute in them together (Liker, 2003), which consequently leads to higher cross-functionality in the company (Söderberg and Alfredson, 2009). Additionally, the Obeya’s walls are covered by different types of data to help the project team make more informed decisions through simple and instant access to required information in real time. According to Liker (2003),

Obeya's purpose includes simplifying information management and on-the-spot decision-making.

Obeya accelerates the decision making process by facilitating communications, sharing information, integrating teams and maintaining alignments (Andersson and Bellgran, 2009, Liker, 2003). Olausson and Berggren (2010) state that Obeya allows managers to prioritize tasks and be more involved in the decision-making process i.e. higher level of decentralization.

Oppenheim (2004) describes Obeya as a program room where Value Stream Mapping (VSM), integrated activities and events, program notes, takt times and lean product development flow are demonstrated on the walls and separate meetings are linked to the program room. Likewise, smaller rooms are required nearby for breakaway discussions, whereas the main room should consist of facilities including networked computers, printers, projectors, ample writing materials and a large conference table with chairs to accommodate the team members. Furthermore, Andersson Schaeffer (2011) mentions that interior design and room size can influence the room's communication and efficiency.

Obeya can indicate different ongoing projects in the company to control the progress and support further efforts (Söderberg and Alfredson, 2009). Andersson and Bellgran (2009) describe some companies that are using a room similar to Obeya to check similar development projects. The aim of such rooms is to integrate product development and production to have better and more production-friendly products. Those companies said the following were benefits for such a room:

- Efficient communication, cross-functional work and accurate decision-making
- Empowering the project identity
- Facilitating the project manager's responsibilities
- Time reduction in information flow by visualization
- The possibility to live and breathe the project
- Shorter lead times for development projects
- Positive inspiration for the company and its employees
- Achieving and absorbing knowledge
- Reinforce the impression of professionalism

Andersson and Bellgran (2009) conclude that:

> *"Obeya can be used to reduce waste in the form of unused creativity, long PDCA cycles, low motivation, complicated communication, low dedication or weak representation of the company identity."*

Obeya's benefits, according to their study, are:

- Helping to make the plan, do, check and action (PDCA) cycle shorter by gathering decision makers in a single place
- Facilitating communication between team members through face-to-face daily contact
- Supporting product development through effective communication and proper technology
- Providing an infrastructure for idea generation and development for new products and cost reduction.

Information sharing among people with different backgrounds and knowledge is problematic. In production system design, different functions need to cooperate; thus, information comprehension and use by all is considered essential. People with similar knowledge and backgrounds can understand each other much easier. Through Obeya, team members achieve common understanding of project procedures and progress, necessary information and what other functional team are occupied with in an easy and quick manner (Söderberg and Alfredson, 2009, Osono, 2008). Furthermore, having different processes and projects in parallel might cause difficulties in information handling. In production systems, the empirical findings of Bruch (2012) determined that high-quality information assists designing and supports justifying each decision even though there is not a specific approach to achieve it. Hence, inadequate information quality causes confusion and delays. It is also said that concurrent engineering increases risks of information overlap, which affects information quality.

Obeya, as a powerful tool to capture and maintain information from different projects and processes, reduces the risk of information affection, since a variety of engineers and managers participate, continuously review and validate the information to check for bias or error. In addition, the applied visualization tools of Obeya help reduce information loss and misinterpretations. According to Bruch (2012), obtaining, sharing and utilizing information continuously requires the entire process of production system design. Obeya is a place where managers and engineers can get together in the production system design process to constantly update each other's experience, information and knowledge while documenting these activities. Obeya is also capable of facilitating personal interaction and communication, which accelerates information sharing through visualization, clarifies undefined or unclear issues, and supports simplified consensus and approval. Obeya enables information sharing inside and outside the project team leading to more data exchange and multifunctional discussions.

Bruch (2012) explains that information availability on digital networks like intranets is not enough, as it might not be accessible for all members. It is not always clear where to find the correct data and it is not possible to have all production system design information available, for instance, late decisions may not be known to all members.

Information visualized in Obeya varies from design graphics, manpower charts, quality information or financial status to other crucial performance indicators related to a project (Liker, 2003, Andersson Schaeffer, 2011).

In this study, we categorized the visualized information in Obeya into six groups:

Table 1- Categorized information in Obeya

Group	Examples
Product specifications	- Project goals and main characteristics - Drawings - Virtual prototypes
Project plan and schedule	- Project plan and Gantt chart (e.g. Microsoft project or primavera) - Project milestones and deadlines - Alert board (e.g. delayed critical tasks or incoming deadlines)
Project organization	- Project organization chart - Team members and assigned tasks and responsibilities
Resource management	- Budget control - Man, machine and other resource availability and schedule - Resource bottlenecks highlights
Idea development tools	- Brain storming board - Tools for capturing results for further uses (e.g. camera for taking a picture etc.) - Brain writing tool (we will talk about it for more description)
Meetings and reviews	- Video/audio/image capturing and playing tools - Minute of meeting of daily reviews and other meetings

The above information can be reviewed by any involved member resulting in any deviation from the standard, schedule or performance goals appear as soon as possible.

Documentation and communication

Bruch (2012) states:

> "*In order to accomplish the interdependent work activities, design information was shared among the representatives from the various functions involved in the design of the production system*".

Information sharing can be performed through written documents (digital or physical) or communication (face-to-face or digital). Communication resolves complicated affairs through faster feedback and confirmation (Bruch, 2012). Written documents are more proper to receive customer feedback (Bruch, 2012). Obeya is capable to handle both kinds of information sharing. Written documents are mostly in A3 report form, a standardized format for information. Reports are then posted on the walls and boards (Söderberg and Alfredson, 2009, Osono, 2008). According to Söderberg and Alfredson (2009), the A3 reports are not fully comprehensible without Obeya room. In Obeya, involved people talk about issues in the reports and untangle the problems. Consequently, it is easier to find the subject while looking at an A3 report, listening and watching the facilitator or responsible person. Further, Andersson and Bellgran (2009) confirmed that standing beside the visualizations on the wall, discussing and indicating make it easier and faster to grasp the project.

As mentioned before, documentation of prior mistakes is essential for companies during their learning process for future projects. As stated by Bruch (2012), documentation and information collected regarding functions, properties and capabilities of the technical, material supply, human and control subsystems is vital to design production systems as well. However, the research of Söderberg and Alfredson (2009) and Bruch (2012) indicate many companies do not concentrate adequately on documentation. In production development, there is a lack of information prioritization and documentation related to the above areas in the early phase of concept generation in industrialization projects (Bruch, 2012). This information is only transferred orally throughout the meetings, without any standard or routine. Therefore, these experiences and information are not established in an inclusive, standardized written format whereas, Bruch (2012) declared that the formalization level is vital for information sharing. When a product is developed, various technical methods are tested to find the right solution. The process of evaluating and testing these options (i.e. why those ideas fail and why this particular one was accepted) need to be documented to use in similar future projects. Optimistically, only the successful ideas are documented while in future projects, wrong ideas might be evaluated again, which lead to time and cost waste. Generally, when the problem is solved it is moved out from the board and system. If there is no system to recall the knowledge about the problem, there is a risk of losing the information and knowledge. As a consequence, it is possible the same mistake will be repeated in the future, which comprise both monetary and time waste for the company. A, high personnel turnover rate between projects or among companies makes this worse because experiences are not documented and those involved may not be available anymore. On the other hand, perhaps the failed idea in one project is the right idea for another project, justifying the importance of documentation. As Bruch (2012) states, replacing production systems become more complex with the absence of information from previous production systems. Thus, regarding the requirement of a system to store A3 reports from a variety of projects (Söderberg and Alfredson, 2009), Obeya is the best tool to collect, save and reuse information and experiences in a standardized format. It also allows for simplistic retrieval in a visualized format in the future for further development or improvement.

Information sharing, in term of communication, can also be conducted through face-to-face meetings or installing digital equipment in the room to communicate with people in other physical places, especially for projects like global product development.

Lean production

During the last decade, globalization has increased and various companies around the world are trying to dominate parts of the global market. But what makes some companies successful and different from others is their competitiveness and flexibility. Now, lean production systems are the most valuable tool to achieve these capabilities, as they have a huge impact on the production process (Liker, 2003).

Less than 20 years ago, the book *The Machine that Changed the World* was published and informed western manufacturers that Toyota is far more advanced than any company in the world for producing cars more efficiently with less time and energy cost and higher quality (Söderberg and Alfredson, 2009; Kennedy, 2003; Womack and Jones, 1991). Although many companies attempted to copy and apply Toyota's system, they could not and still have not reached Toyota's levels of efficiency (Söderberg and Alfredson, 2009; Liker, 2003). Nevertheless, some scholars distinguish Toyota's production system from others with just their intensive tendency to continuously improve solutions (Warnecke and Hüser, 1995).

Toyota has utilized a lean production system as a revolutionary approach, which is known for Toyota Production System (TPS). Toyota's exclusive approach is applicable to any organization or process, no matter if it is in manufacturing or service. The key to Toyota's success is operational excellence, a basic part of lean tools such as just-in-time (JIT), Kaizen, Kaikaku, Jidoka, Heijunka and one-piece flow. These techniques, along with Six Sigma, are the implementation tools for lean production although lean thinking plays the most substantial role in this philosophy (Liker, 2003).

Lean manufacturing is a philosophy concentrating on adding value to customers by eliminating waste and ineffective work. However, value is defined as any process that costumers favorably accept to pay for. Lean is combination of various tools such as Kanban, 5S, JIT, etc. with the goal of increasing efficiency based on flow optimization. To put differently, manufacturing industries have understood that lean is not a couple of techniques or mentioned tools, but a philosophy that ought to be grasped by people involved in the organization or company (Andersson Schaeffer, 2011). Therefore, spreading lean thinking between workers, employees, managers and suppliers is an ultimate goal to maintain a learning organization (Liker, 2003). Besides, lean product development tries to broaden lean thinking through clarifying responsibilities, standardization learning mentality and visualizing data (Söderberg and Alfredson, 2009).

According to Womack and Jones (1991), lean manufacturing can be divided into five levels, consisting of customer value, the value stream, flow, pulling from customers and perfection. Lean production yield for concentrating on flows by adding value in each step, without any stoppage, and pull customer demand to replenish regarding the next station requirement while the whole system should comprise continuous improvement and learning. Further, changeovers, standardization and error proofing are the essential to flow concentration (Womack and Jones, 1996). In other words, lean can be defined as reducing lead time by eliminating waste leading to high quality, low cost, high safety and high spirit within the company from product design to delivery (Womack and Jones, 1991). However, Warnecke and Hüser (1995) defined lean as a system of methods and measurements that lead to competitiveness not just in manufacturing, but throughout the whole company. According to Warnecke and Hüser, the four individual aspects in lean production are product development, supply chain, shop floor management and after-sales service (Warnecke and Hüser, 1995).

Surprisingly, lean production has fundamentals in mass production and removing non-value added work-in-process, comes from scientific management in Henry Ford's era, the 1920's (Andersson Schaeffer, 2011, Liker, 2003).

Sweden, as a country with a large number of high tech production industries, particularly manufacturing cars and trucks including Volvo, SAAB and Scania, has invested significantly in lean production systems within the last decade and lean implementation is visible in a majority of big companies with worldwide reputations. As a result, the Swedish industrial research institute, Swerea IVF, started a project severa; years ago to identify what lean production is and what benefits are achieved through implementation. Several companies were interested in the lean approach got involved and were financed by Vinnova.

Furthermore, various research and investigations have been complete to clarify a vision of lean process to reduce the gap between Japanese and Swedish industries.

Consequently, companies involved in this thesis are chosen among lean manufacturing companies in Sweden, which are at top of the ladder of their proficiency.

According to Liker categorization, waste is divided into seven or eight non-value-adding categories that the companies in this thesis continuously try to remove through morning meetings in their meeting places. In those meetings, problems are shared and discussed with experts from different segments of company to eliminate them. The waste includes overproduction, waiting, unnecessary transport, over processing, excess inventory, defects and unused employee activities (Liker, 2003). However, each company has its own way to hold meetings with the sense of time and frequency of meetings.

Trying to eliminate waste, improving communication among different segments, brainstorming to solve problems, and etc. can be complete in such meeting spaces or Obeya room if some features of Obeya definition are disregarded.

For the first time, Obeya was used to develop the Prius. Experts from different Toyota departments got together in a big room to review project progress, discuss cross-functional issues and make decisions on the spot. Since then, it has become a product development process within Toyota. Obeya gets help from visualization tools to show information clearly to accelerate decision making (Liker, 2003).

The purpose of Obeya is to ensure project success and shorten the PDCA cycle (Plan-Do-Check-Act cycle) by improving information management, reduce production time and bureaucracy, increase team integration, make problems visible and better understand project goals.

Kaizen

Global competition among manufacturing companies requires companies to deal with fluctuations in product demand, volume, variation, life cycle and technology. As a consequence, to stay competitive, continuous process and production development is the most significant requirement (Yamamoto and Bellgran, 2009). Kaizen, Japanese for continuous improvement refers to an incremental process improvement approach to enhance production performance (Liker, 2003). The international research community considers Kaizen the main reason of high competitiveness in Japanese manufacturing companies (Yamamoto, 2010). Whether a small or big improvement, it should be incessant and follow the lean philosophy and Toyota Production System which led to organizational learning (Liker, 2003, Imai, 1986). However, it does not mean that only Toyota performs it. Kaizen has become a characteristic at top companies in Japan. This concept is well recognized by most companies around the world and various tools and techniques, including quality control circle, policy development, Mottainai and Muda, have been developed and are being applied supporting Kaizen (Yamamoto, 2010, Liker, 2003).

Kaizen consists of an effective group working, problem solving, data collecting, analyzing, documenting and improving processes. Kaizen is based on Deming cycle or PDCA cycle (Liker, 2003). In another way, Kaizen can be defined in three steps: see, think and act. In the first step, the shop floor is carefully observed with severe eyes to identify problems. In the next step, a solution is defined and implemented (Yamamoto, 2010). Additionally, Kaizen can be considered in three perspectives including process orientation, people orientation and small-step improvement. Process orientation refers to process development while good results automatically appear. People oriented means all involved from top manager to bottom-level employee should believe in continuous improvement (Berger, 1997; Imai, 1986). Similarly, Brunet and New (2003) defined the same features for Kaizen including continuous, incremental and participative. All activities are implemented by holding decision-making sessions and obtaining group consensus (Brunet and New, 2003; Liker, 2003)

Moreover, whenever waste appeared, all employees have to eliminate it using Kaizen. One technique within Kaizen is 5whys to get to a deeper level of the problem. This means when a waste is exposed, the cause of the problem should be found by asking “why” until the main reason become clear (Liker, 2003). According to Yuji, severe eyes or Gemba eyes are crucial in Kaizen to find any problem at the shop floor, to get a holistic perspective of operation management and keep improving the shop floor. Furthermore, improvement can be performed using simple but creative methods, such as pen and cartons (Yamamoto, 2010). Additionally, the time between recognizing the problem and taking action is important. Sometimes, regardless of risks and uncertainties, improvements ought to be initiated to obtain further and better ideas even with small changes. With the perspective of performance enhancement and organizational learning, faster reactions lead to a learning by doing approach (Yamamoto, 2010).

On the other hand, Kaizen is not achievable until the process is standardized. First, it should be standardized and stabilized to implement Kaizen. Consequently, typical activities become part of the standard process and evenly decrease waste (Imai, 1986).

According to Yamamoto, to engage Kaizen in industries, eight guidelines have been developed (Yamamoto, 2010):

Problem discovery by severe eyes:

1) Observe a shop floor with severe and critical eyes in perspective of 5S, visual control, 7 Muda and so on.
2) Never be satisfied with current operation
3) Repeat why when one sees abnormalities
4) Do not blame operators, but blame system or standard

Generating simple but creative solutions:

5) Use wisdom thoroughly before using money, use simple handmade devices or solutions
6) Create temporary solutions even if the optimal solution is unknown or takes time to be implemented

Taking immediate actions:

7) Initiate change right away when a solution can be implemented immediately. Small improvement can be performed on the spot especially 5S problems.
8) Initiate change even if there is an uncertainty. More improvements will be found after the change. Small changes can be undone simply enough.

As mentioned before, Kaizen has been well established among scholars, managers and practitioners. But it is still difficult to implement Kaizen within industries even though Japanese manufacturing companies, such as Toyota, comprehend and utilize the concept. As a result, despite a well conceptualized definition of Kaizen in the literature, practice of Kaizen is not well performed within manufacturing industries (Liker, 2003, Yamamoto, 2010, Yamamoto and Bellgran, 2009).

Obeya, as a lean tool in manufacturing industries, can accelerate Kaizen implementation. Within Obeya, other lean tools can be considered and questioned if they are implemented to a satisfactory degree. Combination of Obeya and Kaizen can be referred to Kaizen corner, which was built for academia, short daily meetings with managers and group leaders and engaging laborers through training and morning meetings (Liker, 2003). Kaizen corner integration at case study B is described in the next chapters.

Kaikaku

Kaikaku is a Japanese translation of drastic improvement (Yamamoto, 2010) and refers to production improvement in a frequent, radical way in comparison with Kaizen which is small, gradual and continuous. Kaikaku follows a top-down approach since existing practices are fundamentally changed within a knowledge, technology, method or strategy. Kaikaku's results are higher than Kaizen, between 30 percent to 50 percent according to Yamamoto (2010). However, in other literature, Kaikaku has been defined as "Kaizen blitz," a profound improvement in a limited area and limited time duration, performed by small group of people (Bicheno, 2004). By any definition, the best example of Kaikaku can be found in mobile phones. Competition among phone manufacturers started in late 1990 when nearly all mobile phone manufacturers combined the functions of a personal digital assistant (PDA) with mobile phones. Then, during a very short time from 2000 to 2005, new functions were added to mobile phones that allowed users to listen to audio, take pictures, capture video, use touchscreen display, read office files and hundreds of additional functions. All of these additional functions can be considered Kaizen. But suddenly in 2007, Apple Inc. introduced their first mobile phone with a multi-touch screen instead of the typical smart phones different keypads and displays. Apple dramatically changed the markets requirement and caused all mobile phone manufacturers to change their direction toward this new requirement. At that point, mobile phone manufacturers needed a huge change to focus on innovations and new ways of competitiveness. Kaikaku happened in most of them by putting away the traditional mobile phones and starting to produce new smart phones that can answer the new market requirements.

It should be mentioned that Kaizen is considered part of Kaikaku, despite Kaizen's utilization to maintain improvements of implementing Kaikaku (Imai, 1986).

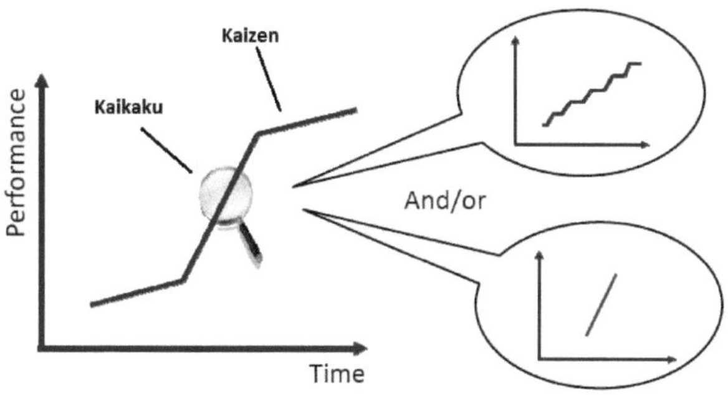

Figure 3 - Kaizen and Kaikaku (Yamamoto, 2010)

According to previous literature studies, companies that are capable of implementing continuous improvement and Kaizen during daily production are better prepared to handle Kaikaku projects and sustain their competitiveness. Brunet and New (2003) assert that Kaizen makes it easy to conduct Kaikaku projects within a company by preparing employees' minds for radical changes. Likewise, as the people of an organization are vital and the most value (Liker, 2003), companies who prepare mindsets of workers, inventors and managers in earlier stages, are more successful than those who only focus on technological aspects (Dilokpimol and Surasawadi, 2012). Besides, radical improvement needs change and readiness in all areas, not just production system. Changes in workplace, involved people, organizational structure, etc. is essential to support Kaikaku.

Furthermore, Yamamoto's model (Figure 4) helps comprehend Kaikaku by dividing it into four different types that occur within a company.

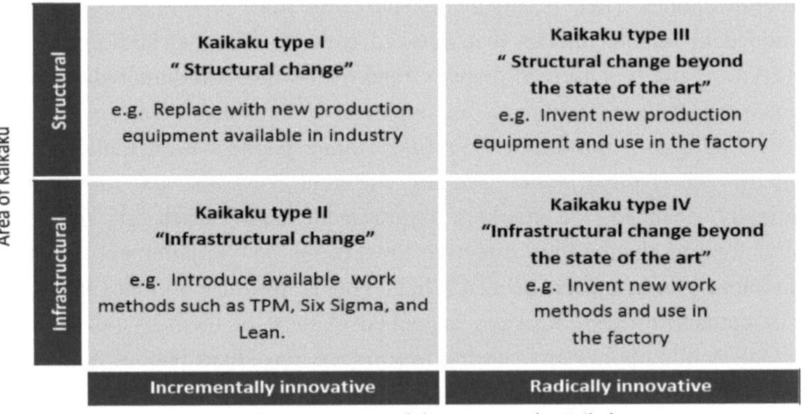

Figure 4 - Categorized Kaikaku (Yamamoto, 2010)

Kaikaku can be new for a company or entirely new for industry. As a result, the horizontal axis of the above model consists of:

Incrementally innovative: When a new production system is the result of Kaikaku while the system already exists, i.e. it is new for the company but not for the industry. For example, implementing Six Sigma or lean or even 5S.

Radically innovative: When a new system is introduced to company as well as industry (i.e. it is new for both).

Area of changes can be also different, so we have two types:

Structural: These changes are more basic and often hard to undo, like automation, production capacity and volume per year.

Infrastructural: These changes happen continuously and need improvement during the implementation such as cost control and maintenance.

However, when and what type of Kaikaku is suitable for company should also be considered. It is said that infrastructural changes should be performed first and operation and layout improvements should be done before considering new equipment or automation. As mentioned before, incremental Kaikaku is also required before radical Kaikaku. The best practice of Kaikaku is to implement according to the sequence:

1. Kaikaku type II, "infrastructural change" like TPM.
2. Kaikaku type IV, or I, which are respectively infrastructural change beyond state-of-the-art and structural change, depending on the company's conditions.
3. Kaikaku type III, structural change beyond the state-of-the-art.

The most important part is how Obeya would help Kaikaku. According to this categorization, Obeya, as an existing tool, cannot be placed in radically innovative Kaikaku. On the other hand, the Obeya concept can be placed in infrastructural change since it tends to have long-term impact, it is difficult to undo, requires investment and works like a facility or an equipment. As a result, Obeya fits into Kaikaku type II, structural change. Fortunately, Kaikaku type II is the most common Kaikaku among companies which have experienced it, although it is the most common, a majority of companies prefer not to report it to the public. Nevertheless, respecting the trend of lean transformation in companies around the world, especially in Sweden, the need for Kaikaku activities cannot be neglected and consequently, Obeya.

Visual management

Visual management tools, along with computer-assisted design, are utilized in Obeya to support the meetings and consensus. Visual management tools help the team check schedules, checkpoints and different aspects of the project in the shortest possible time (Liker, 2003). In other words, visual management is the fundamental tool in Obeya that are demonstrates increased communication and information sharing (Andersson and Bellgran, 2009). Obeya is capable of holding stand up meetings in mornings and sit down meeting in afternoon. Activities and critical issues that ought to be addressed during stand up morning meetings including production goals, resource assignments, previous problems and quality, delivery and cost difficulties. On the other hand, sit down meetings are held after shifts following to discuss delivery status, production disruptions, scrap parts and defects and safety issues resulting in action lists for the next day or shift. Nonetheless, constant meetings help team members achieve higher levels of efficient information (Bruch, 2012).

Visualization and attractiveness of common space enhances the "proximity or exposure to one another," which provokes face-to-face communication (Andersson and Bellgran, 2009). Physical proximity increases information exchange by enhancing frequent information sharing in an informal, face-to-face contact (Bruch, 2012) and facilitating mutual understanding (Bruch, 2012). In addition, informal communication, even in halls, at lunch or coffee breaks leads to better collaboration in process and less uncertainty. As mentioned above, by gathering involved engineers and managers in a single room and illustrating different information, the requirements of walking around has disappeared. Therefore, distance is shortened and walls around management are broken between involved people to have a higher level of cooperation and on-the-spot decision-making. Furthermore, in production process and development, communication plays an indispensable role controlling information transfer (Bruch, 2012). During the production system design, information regarding problems and challenges is shared and risk of delays, lost and changes are reduced. Communication allows quicker feedback, discussions, elucidation and issue solving. Hence, Obeya provides a place for frequent face-to-face meetings and facilitates communication among the involved people.

Obeya is a capable tool in supporting Kaizen and eliminating the eight wastes in production (see chapter 2). Illustrating different information throughout the company makes sales data, costs and quality visible, enhancing involvement. Visualization tools in Obeya localizes problems immediately (Andersson Schaeffer, 2011). Moreover, Obeya leads to production improvement and Kaikaku in industries (Andersson Schaeffer, 2011).

Visualization is a strong tool in lean implementation as one fourteen principles of Toyota, "Use Visual Control So No Problems Are Hidden" (Liker, 2004). This makes the process transparent and enables involved people to see the various process aspects and its situation at any time. Therefore, quick status feedback is given to fulfill requirements (Parry and Turner, 2006; Andersson and Bellgran, 2009; Womack and Jones, 1996). Visualization implies any communication tool that helps team members understand immediately whether the work is performed in a correct way or if it deviates from the standard. It indicates the exact status of the process with various vital workflow information certifying fast and suitable operation and process execution (Andersson and Bellgran, 2009; Liker, 2003). Visual management guarantees the organization's goals are apparent and all essential information is accessible to complete the work as efficiently as possible. Visualization helps compact data into a visible format that is controllable, manageable and easy to discuss (Sobek and Smalley, 2008). It assures communication and increases knowledge sharing among involved people (Lindlof and Soderberg, 2011).

Visualization systems increase productivity and safety; decrease mistakes, defects and costs; facilitate communication; leverage investment and resources; and provide greater control over work environments (Liker, 2003). Furthermore, visualization helps reduce the eight wastes by supporting fruitful communication, cross-functional work and decision-making procedure; strengthening the project identity; facilitating project management; reducing lead time for development projects; and inspiring workers in a positive way (Andersson and Bellgran, 2009). The best visualizations consist of graphical representations, posters, pictures, symbols, transparencies and color-coding along with audio signals (Bilalis et al., 2002). The human brain processes images easier than text. Research indicates that humans automatically pay attention to mobility and visual objects (Andersson and Bellgran, 2009). It is believed that within the Toyota system, visual management complements humans, as the human brain is audibly, tactilely and visually oriented (Liker, 2003).

Since a common problem among managers and engineers is information overload, sharing knowledge and information in a visual approach leads to simpler, quicker and more effective results than reading long documents full of text (Parry and Turner, 2006; Martin and Remo, 2007). The most time consuming way to express ideas is to present them in a lengthy report with technical descriptions and data. A variety of methods is utilized to help this principle smooth product development and production processes. The most important tools in visual management are the meeting rooms, A3 reports, whiteboards, Andon boards, updated standard work charts, etc. (Andersson Schaeffer, 2011; Liker, 2003; Parry and Turner, 2006; Oppenheim, 2004; Sobek and Smalley, 2008).

It should be mentioned that this section does not attempt to explain or review all the visualization tools in the lean practitioner's package. The concentration is on visual control tools inside the Obeya/meeting spaces. Other tools, such as Kanban, the one-piece-flow, 5S and shadow paint of tools, associated with lean production to improve and facilitate the flow to make any standard deviations visible are not mentioned in this research.

By visualization, team works are affected and improved as project teams obtain higher levels of knowledge transfer within the projects (Lindlof and Soderberg, 2011). In the current literature, knowledge is separated into two types: tacit knowledge and explicit knowledge (Söderberg and Alfredson, 2009). Explicit knowledge is "knowing about" (i.e. facts and theories), while tacit knowledge is "knowing how" (i.e. how to do things). However, transferability is the main distinction between them. Explicit knowledge can simply be documented and spread in an organization through reports and illustration while tacit knowledge is achieved through practice and transfer is often expensive, uncertain and time consuming

(Söderberg and Alfredson, 2009; Jugdev et al., 2011). It supports and promotes tacit knowledge transfer in terms of quality and speed (Lindlof and Soderberg, 2011; Martin and Remo, 2007).

This information ought to be displayed in a way that adds value to the process. Common information in production areas includes production output rate, safety issues, defect rates, quality information, human resource leveling, activity priority, delivery schedule and production system. To manage this information about different activities in various projects and processes across different divisions, a usual approach is utilizing software-base tools (Liker, 2003). One goals of information technology and automation is to make offices and factories paperless. The internet and computers make it possible to gather large amounts of information, both in written and visual forms, quickly and share them through email or software.

Despite the above requirements, most companies still use physical rather than digital visualization. The answer lies on usage of digital visualization. Extravagant use of software leads to local and relational differentiation and increases the requirement of information processing capabilities (Lindlof and Soderberg, 2011). It means less communication among engineers and managers, which is counter to lean principles in some aspects. Furthermore, physical boards may be easier to display and review data in comparison with software-based systems (Lindlof and Soderberg, 2011; Parry and Turner, 2006). According to Parry and Turner (2006), limitations of the physical visualization boards allow more concentration on the quality of information illustrated. Likewise, software or computer-based systems inevitably create operator experts, though cause inherent drawbacks. Moreover, looking at screens creates isolation and removes attention from where the real work is being done (Liker, 2003). By using electronic boards, a few employees take control which leads to loss of group ownership. In contrast, all team members are able to change the physical boards, allowing equal authority among members. Although electronic data can be projected and illustrated on screens, they cost more both to be purchased and maintained (Parry and Turner, 2006).

Contrary to the above statements, the disadvantages of physical boards and notes are:

- It is more difficult and time consuming to update the current situation precisely because the boards and sticky notes are artifacts
- As artifacts, the boards and stick notes are not documented for saving historical data related to project progress or for later follow-up and knowledge sharing
- As soon as a note is moved or erased, any link ceases to exist anymore to show how the moved task would have impacted the other parts of the project or team members work

- Amount of documentation in practice is limited
- Due to multiple sites, it is difficult to connect the boards to each other
- Connecting causal link among activities is problematic

All above disadvantages are expected to be solved using digital or software-based visualization technology (Lindlof and Soderberg, 2011). Oppenheim (2004) asserts, "The wall layout is preferred to electronic implementation, because it enables the core team members to read all tasks, brainstorm and negotiate in real time the task parsing, precedence, concurrency, synchronicity, scope and effort, inputs and outputs, and waste to finally reach a consensus." However, unlimited document saving may lead to overwhelm, confuse and probably prevent reviewing and controlling information (Parry and Turner, 2006; Taxén and Lilliesköld, 2008; Bruch, 2012). Nevertheless, avoiding information technology is impossible and no one can hide from it. The challenge is to find a balance using available technology in the best way to support employees by improving visual control. This balance may be a combination of both approaches as Toyota uses a wall-sized screen to show 3D images of vehicles in the service parts warehouse in Hebron along with physical visual signals (Liker, 2003).

As mentioned previously, boards are one of the most common and simple tools in visualization (Andersson and Bellgran, 2009) that is used in various environments especially offices and floor shops. Standing beside a visualization tool-like board and looking at it, discussing it and showing different aspects (Andersson and Bellgran, 2009) helps enforce a continuous work flow by displaying the operation status in a determined period or pace of operation, progress of process and takt time (Liker, 2003). Consequently, the visibility of each tasks progress is immediately indicated to provide extra effort. In some companies, "dashboards" demonstrate information about the current situation of service, production and process with graphical outputs and performance indicators (Parry and Turner, 2006). Furthermore, board placement is crucial as it supports interaction and leads to exchange planning information. According to Liker (2004) and Anderson Schaeffer (2011), the best place to visualize is at the site, which lets operators and production technicians understand the standard deviation. In addition, crucial problems or unclear issues can be attached to the boards to be discussed and elucidated. Boards are used to hold meetings in the center of production to check daily progress, Kaizen activities and review the process state. Absence or presence of labor is also displayed on the magnetic boards to indicate the current availability of individual workers, departments or divisions in the firms (Parry and Turner, 2006). This feature leads to easier workload leveling during the shifts. Boards become more advantageous by using colorful notes to distinguish different issues such as problems, deliveries, defect rates and production ratio in each division

(Lindlof and Soderberg, 2011; Parry and Turner, 2006). Team members at Toyota constantly update visualization white boards called process control boards (Liker, 2003). The purpose of such well-designed boards with charts and graphs is to indicate the current status of the projects and employee workload, inform of plant operations, demonstrate a desirable view of the project and support the project teams to have a better perspective of project progress. Additionally, cameras can be installed to capture the information on boards. Later, the captured information could be displayed on sheets or screens in another division. Some companies go further and use a completely electronic visualization in a room where team members stand around a big board. This big board gives a higher feeling of ownership (Parry and Turner, 2006).

It is often said each team ought to develop its own visual management boards to have full authority and control of them as different teams have different knowledge. The teams may also use different kinds of information with different standards. The level of benefits from using the board can also be different. Thus, various standards, information and levels require higher ownership and "one-size-fits-all" boards need to be avoided because each team, division or department has its own objectives. Having a different board in a different part of firm drives the principles (Parry and Turner, 2006):

- Concentrating on the delivery in a form of time, quality and cost
- Guaranty efficient resource utilization
- Bottleneck and work in process (WIP) identification
- Showing key performance indicators which are affected by activities and focused on Kaizen
- Creating cross functional communication through boards

From another perspective, information availability is not a problem, but information communication is often not efficient enough (Bilalis et al., 2002). Communication is essential to providing new knowledge and solving problems among involved actors (Andersson and Bellgran, 2009; Taxén and Lilliesköld, 2008). Real time data on the boards lead to efficient communication by reducing the quantity of information exchanged via email, paper, reports and discussion, which wasted time (Lindlof and Soderberg, 2011). Frequent face-to-face meetings and cross-functional conversations lead to more information flowing between team members and displaying a holistic view of the project progress (Lindlof and Soderberg, 2011). This communication helps identify problems in earlier phases of the project, level the workload and manage tasks more efficiently. Furthermore, clear communication guaranties the above mentioned information is understood across the company (Parry and Turner,

2006). Visualization boards support the possibility of spontaneous argument about information demonstrated on the boards in daily work (Parry and Turner, 2006; Olausson and Berggren, 2010).

As mentioned, the possibility of problem solving in early phases is another benefit of visualization. This is due to frequency of meetings and visualization tools usage. By visualizing activities, deliveries, problems, etc., people tend to discuss their problems with others. Through visualization tools, managers can prioritize activities easier and engineers get more involved in the decision making process, which implies a higher degree of decentralization (Olausson and Berggren, 2010).

A common tool in Toyota is that managers and engineers put information on an A3 piece of paper. The A3 report is a one sheet standard report documentation process by displaying graphs, charts, text or pictures to maintain and develop the experience. A3 is an easy format understood by all members. It contrasts with written reports, which a lot of information needs to be reviewed to find the right information, causing difficulties. Sharing information with others on the A3 reports leads to higher information quality, receiving correct inputs and achieving consensus. By having inadequate information quality, the inability to find or use the right information would be raised (Bruch, 2012). An A3 report would concisely mention the problem, current status, cause, suggested solutions, chosen solution and cost-benefit analysis (Söderberg and Alfredson, 2009). The A3 report is based on Deming's PDCA Cycle. However, the A3 report starts with a deep understanding of the current situation. Moreover, the A3 report can even be considered a learning process that helps to hold effective meetings in term of quickness and being interesting (Söderberg and Alfredson, 2009).

To design, develop, manage and improve a process, value stream mapping (VSM) is a fundamental technique in lean manufacturing. VSM's objective is to eliminate non-value-adding activities, decrease resources, delete waste and reduce operation time(Parry and Turner, 2006). Another lean tool that can be used in meeting spaces is an Andon board, which is utilized to provide information about production status and its problems.

All these visualization tools are used to communicate data during meetings. These meetings facilitate the communication process by helping employees achieve desired consequences. Meetings are held regularly around the boards to evaluate the project progress in each division. They might be held with different levels of participation and length of time. At the meetings, people provide updates about the status of their division, which facilitates the knowledge-sharing loop (Parry and Turner, 2006). There are several requirements to hold an effective meeting. Without them, the meeting would be pointless, time consuming and ineffective (Liker, 2003):

- Clear meeting objectives with clear tasks and deliverables
- Right people at the meeting
- Each participant should prepare the necessary documents for meeting
- Powerful visualization tools
- Information sharing prior to the meeting
- Determined meeting duration

Meetings are the best place to discuss problems and projects' contents. Group meetings are also considered the most intuitive way to share knowledge (Söderberg and Alfredson, 2009). Some meetings are held on the shop floor to handle daily short tasks in a project. Bigger meetings are held occasionally with project leaders to handle parallel projects. However, some companies use "technical meetings" to obtain a higher level of standardization. Obeya can handle all kinds of meetings to accelerate the decision-making process.

Visualization is an inseparable part of Obeya and is one of the first steps toward lean implementation (Lindlof and Soderberg, 2011). In Obeya, deliveries, activities, outlines, project goals, characteristics, milestones and deadlines can be displayed on a physical and/or digital boards while discussions and brainstorming sessions can be held. Obeya, war rooms or program rooms are a strong visualization tool in lean that illustrates information in a visual, living or/and electronically searchable format. It also gathers engineers and managers to expedite the decision-making process and consensus (Morgan and Liker, 2006a; Andersson and Bellgran, 2009; Söderberg and Alfredson, 2009; Oppenheim, 2004). This room, through the mentioned visualization tools, controls project development, facilitates cross functionality and information accessibility, accelerates consensuses' building and helps understand what the other functional teams are doing (Söderberg and Alfredson, 2009; Morgan and Liker, 2006a). The same concept is occasionally used as Kaizen corner to show information.

Project management

Generally speaking, projects are defined in term of plans, tools, organizations and resources with a specific objective. Project management is about arranging, modifying and constitutes related matters to obtain the goal with some limitations (Taxén and Lilliesköld, 2008). Usually, projects are assessed according to tradeoffs including time, cost and scope. Optimizing each criteria leads to entire project optimization and that's management's goal. The project management process has received more attention over the past last decade, as it is a powerful tool in firms' competitiveness (Jugdev et al., 2011). However, complexity of projects caused managers to holistically view projects and unable them to view project details, including individual work tasks. It is mostly attempting to allocate resources to shorten the activities. However, long-term development in production and product receive less consideration (Söderberg and Alfredson, 2009).

An essential problem in project management is facilitating tacit knowledge sharing. As mentioned previously, tacit knowledge is shared informally through social exchanges and practices (Söderberg and Alfredson, 2009; Jugdev et al., 2011). Obeya, through regularly gathering engineers, managers and workers from different divisions helps spread knowledge by exchanging ideas and communication practices. As a result, in addition to explicit knowledge sharing, it would lead to tacit knowledge sharing among team members. Obeya, through removing the barriers to communication leads to knowing what others do, what problems exits, who is struggling and makes it simpler to discuss problems. Furthermore, Obeya provides a place for participants to prioritize the project (Andersson and Bellgran, 2009). Moreover, coffee breaks or "Fika" can be taken in Obeya, which helps informal knowledge and experience transfer between involved people. Andersson and Bellgran (2009) explain that Obeya affects work organization by increasing the operators' and production technicians' interest to spend more time in the meeting space socializing and gaining technical information. To achieve the above purpose, common tools such as Gantt charts, Program Evaluation and Review Technique (PERT), Critical Path Method (CPM) charts, Work Breakdown Structure (WBS) and network diagrams can be visualized (Taxén and Lilliesköld, 2008) to facilitate the projects management.

Additionally, here are learning processes that Obeya could solve (Söderberg and Alfredson, 2009):

- Post-project appraisal: A group of specialist in an organization after completion of the project examines and analyzes the project to write a report about what has been learned.
- Project Audit: A cross-functional group from outside the project interviews project members across all levels in different areas. Afterward, an audit group analyzes and identifies the weaknesses and strengths of each area. Finally, a recommendation is given for future improvement.
- Post control: Project manager chooses the members and functions involved, resulting in documented future improvements.
- After-action review: 20 minutes to 2 hours to review the project by asking questions including:

1) What was supposed to happen?
2) What actually happened?
3) Why was the cause of differences?

4) What can you learn from this experience?

Results are converted into flip charts to handle similar projects in the future.

A critical issue in project management is to allow project members to get involved and follow the project to gain experience for future projects. This improvement is obtainable from other projects that are being complete in parallel. Consequently, experiences and suggestions help develop the way work is performed.

A white book is a document consisting of projects' experiences. It is primarily written at the end of the projects and includes project mistakes, successes and experiences during the projects. It might be written in different ways, such as bullets points or storytelling, depending on the project manager. It is useful for future projects to learn from previous experiences and take advantage lessons learned. Learnings are mainly written about the problem, actions taken and how the problem was solved. Since it is mostly written at the end of each project, after the typical 2-4 year project time, a lot of information is missed. Moreover, there is a tendency to remember positives rather than negatives. Accessibility and reliability are also big problems for the database. If someone looked for hours and doesn't find the answer, the next time he/she will probably not try it again. In another point of view, personnel turnover rate between projects or among companies leads to losing project experience and knowledge in product and production development. During each project, there is possibility of changing project leader or other members. Consequently, this might cause new objectives, schedules or resources. Therefore, Obeya can help contain all this information to show and present to the proper person and at the appropriate time. Obeya, as a tool to keep current information on different projects, covers all above white book problems. This daily-updated white book can be exchanged between projects through Obeya and its visualization tools regardless of company turnover rate. This tool leads to reliable knowledge sharing between different projects in different divisions through supporting various type of documentation including Micro articles, RECALL/Wiki, white book and A3 reports. The ability to transferring knowledge among different projects enhancing international competitiveness (Söderberg and Alfredson, 2009). Obeya allows all people in the company or organization to change, add or reduce the information. It is forum to communicate; share knowledge and experience while informing others of job progress, who owns specific knowledge, who to talk to about special issues and who to get for help. Regular meetings near the visual boards would increase the project efficiency. In regular meetings, people get involved in different projects, so members get something in common with what others are working with (Söderberg and Alfredson, 2009). In this way, experience and knowledge is moved forward while members are improved and developed (Söderberg and Alfredson, 2009). Furthermore, problems do not store on personal hard drives but on a network available to everyone.

Chapter 4: Empirical studies

In this chapter, information about the case studies is presented. Table 2 summarizes information about the four cases. Afterward, a cross-case analysis is performed to compare the cases and, finally, based on case study results, recommendations are given.

Case studies

In this research, four cases were selected for study to gather information about the current use of Obeya and similar meeting spaces. The studies were limited to the use of the meeting spaces in production system development activities and product development-related activities are not included due to subject irrelevance to the research and also limited access to product development-related activities and data. Cases were selected based on the following criteria:

- Geographical location: All selected cases are located in Sweden. This criteria helps minimize the influence of work culture and other cultural differences. Easy access to production facilities were another positive effect of this criteria. This assures a certain level of quality and working standard regarding Swedish legislation and standards as a leading country in such issues.
- Global footprint: To avoid negative effects of all cases being from a single country, a broad and considerable global footprint of the companies was established as another criteria. In this regard, all selected manufacturing facilities are a part of an international company leading their respective industry. This criteria ensures the selected cases consider the global, in addition to local, requirements and standards. Also, it is an indicating factor of the position of the company in the market and its involvement in the highest level of global competition which, in turn, assures the validity of the results in this aspect.
- Lean production application: All selected cases use lean production system and its related tools and methods. Each has been practicing it for at least four years. This criteria is critical because it is necessary to have the same general frame of process and development in all the cases. As Obeya and similar meeting spaces are an integrated part of lean a production system, they cannot be studied out of context which is the lean production system. In this regard, all the selected companies are committed to implementing lean production systems while each case has its own customized interpretation system according to its requirements and specifications.

- Company size: to capture the effect of the company's in their uses of meeting spaces in production development activities, two cases were selected from medium-size companies and two from large companies. Small-size companies were excluded because the high level of customization in their processes and lack of global footprint in most cases. Diversity in size allowed the authors to investigate the practice of using Obeya in different companies with different sizes and compare differences according to their characteristics.
- Product: While all cases were selected from manufacturing industries, authors kept the diversity of the companies regarding their products. This criteria was established to avoid neglecting aspects of the research because the special nature of some products or their respective production system. Level of automation, product size and production rates are some aspects covered in the selected cases.

Case A: BT trucks

Company description

BT is a leading manufacturer in lean production and part of a leading international group in the material handling equipment industry with 1,800 employees. Their annual production volume is 52,000 units and nearly all the products are sold Europe. Daily production capacity is 250 and there are 250 product variations which all are produced in the same production and assembly lines. The production facility consists of five assembly lines, a welding and a painting department where all eight product groups of the company are produced. These product groups are hand pallet trucks, pallet trucks, power stackers, reach trucks, counter-balanced electric BT cargo vehicles, counterbalanced IC BT cargo vehicles, order picking trucks and very narrow aisle trucks. The interviewee with the company was a production engineer and had been in the company for approximately six years.

The company is enjoying lean production system and TPS since it was acquired Toyota approximately 15 years ago. As a result, people follow Kaizen as part of their daily work. On the shop floor level, a considerable part of Kaizen activities are involved with improving developing production processes as a part of production system.

Meeting spaces

A single meeting place is used in BT trucks to manage daily production problems and Kaizen projects. Every line has its separate six-minute morning meeting and additional meetings are held to discuss problems before the meeting and defects of the related line with responsible people in attendance.

Problems are described briefly in the meeting and documented in a pre-defined form. Responsible people are assigned to the problem solving process according to the nature of the problem. For instance, if the problem is caused by one welding station, the welding department supervisor, the station operator, the process customer (out of the department, if required) and responsible maintenance technician form a team to find the cause and solution. A detailed follow-up of the problem and solution remains for a meeting between assigned persons. The semi-filled form stays on the board for follow-up.

The usual process includes 24-hours for finding the cause and suggesting a solution under the supervision of the related line or department supervisor. A specific section is assigned the 5Whys analysis in the pre-defined form. After developing the solution, it is documented in the respective part of the form and discussed briefly in the meeting the next day. Finally, the completed form is placed on the board as a lesson learned and for documentation.

There are specific tools used in the meeting space, including pre-defined A4 forms on the board for registering the problem. The general thought behind these forms are the same as Toyota's A3 reports, but they are more structured and utilize a template. The structure and template was developed over the course of time from repetitive and common characteristics of problems in each department. In this regard, the departments newer to the process use less structured or unstructured forms, which in some cases is a blank A4 paper. Pictures or other visual tools, like charts and diagrams, are used in the reports to simplify understanding and facilitate solution development.

The forms remain on the board in the area of the department and each department has its own space for documenting and discussing problems, but the board is in a room available to people from other departments. This feature helps facilitating data sharing among different departments and daily meetings help transfer knowledge generated throughout the factory. Along with mentioned A4 forms, whiteboards are also used to support instant visualization and ideation.

Data registration and visualization processes are manual and there is no indication of using digital tools in the environment. According to the interviewee's answers, the process has remained manual to avoid disadvantages of digital documentation such as easy modifiability, less user interaction and risk of data loss. In this regard, all forms are filled by hand and kept on the room walls on the assigned space to each group as a visualization tool for follow-up and periodically archived to free space for new documents.

Interview details

The interviewee at BT in Mjölby was a customer relations manager and had sufficient knowledge and experience in material handling production to guide the authors and answer questions. She had been working for BT for more than 10 years and, not surprisingly, was the only interviewee who knew about Obeya and its advantages. As indicated in the research approach, a semi-structured interview was performed at BT to be flexible and let the authors create new questions and ideas throughout the interview. The interview took about two hours, including visiting the production line and meeting place. However, during the visitation, questions regarding Obeya and research objectives were asked, recorded and documented. Pre-defined general questions were formulated according to thesis objectives, research questions and literature study that had been completed previously. These questions were about the meeting place and its use, the number of participants and their positions, duration of meetings and information visualized in the room.

Case B: Volvo GTO in Köping

Company description

Volvo Köping is large company with around 1,300 employees. It is part of international automotive manufacturer, Volvo Group, the biggest manufacturer in Europe and the second leading world manufacturer in their business area, with different global brands such as Volvo trucks, Renault trucks and Mack trucks. Volvo Group Technology Organization in Köping manufactures heavy-duty engines, gearboxes and drive shafts.

The production factory studied has three main types of gearboxes including those for trucks and busses, construction equipment and the marine drivelines. Gearboxes for trucks and busses is the largest part of production. These gearboxes are manufactured in a number of different variants, and annual capacity is approximately 132,000 products. In addition to production and assembly lines for each of the three main production areas, there is also a maintenance department supporting all the lines.

Meeting spaces

Each production area has its own meeting place but there is general design for meeting places in the production and assembly departments. This design includes a meeting corner near the related line and a visualization board showing information about problems, ongoing Kaizen projects, quality measures and safety issues. The maintenance department has it is own special room design, referred to as Kaizen Corner. There are a number of customized visualization tools including different schedules and reports for ongoing maintenance projects.

Generally the same process as company A is followed in company B. Short meetings are held every morning with main actions to follow up problems from the previous day. In Case B, contrasting with case A, the meeting has an agenda regarding the problem category: quality and product defects, problems in production system and safety issues.

To follow up the problem, nearly the same process as case A is followed to find the cause and solution. A team is assigned to problem solve and to follow up in more detail. The main difference is that no formal deadline is established for the team for a solution. However, the general expectation is to complete the process in a week.

Again, the thought behind reporting in the meeting spaces is Toyota A3 reports and report templates are used for production and assembly lines in the production facility.

Since every department has its own meeting place, customized visualization tools can be introduced according to their needs including tools used in the maintenance department. In production and assembly lines reports, templates are used for documenting and following up Kaizen projects. But in the maintenance department, different and more sophisticated tools are used according to the needs of this department. Examples include resource charts, simplified project schedules and a traffic light status viewer of project goals. In this department, indications of meeting space use for radical change projects were visible.

There were no digital visualization tools in any of the spaces but the interviewee believed that, "One of the future steps to improve these spaces is digitalizing part of the visualization tools" to increase efficiency and ease of knowledge sharing and using.

Interview details

At Volvo Powertrain in Köping, the interviewee was a management trainee/project manager. He had been working there for almost four years and had good knowledge and experience of production lines and different departments in Volvo. Thus, he was able to answer the authors' questions regarding the objective of research and meeting places/Obeya at Volvo.

A semi-structured interview was conducted at Volvo because of flexibility and discussion during the interview. The interview took approximately 2.5 hours including a Volvo presentation, interview, production lines visit and maintenance department visit because each has their own meeting place. Questions that were asked during the interview were all prepared according to previous literature study and experience from the last interview and case study. The authors tried to cover all aspects of Obeya in the interview to answer the research questions and fulfill the research objective. Pre-defined questions were almost all about meeting place characterizations, meeting duration, visualized information types, discussion types at meeting places, participant numbers and their positions, meeting place purpose and, at last, meeting place advantages and disadvantages. Similar to the previous case study, the interview was recorded and authors notes were documented to not lose points and for future use.

Case C: Gyproc

Company description

Gyproc is a medium size company that produces construction material, and is part of a leading international construction material group. Company C has roughly 360 employees, with production facilities in Scandinavia and around 100 employees in the studied facility.

Company C is one of the lead companies in world class manufacturing (WCM) since 2004. WCM enables them to increase their productivity, environmental improvement and product quality. This "best practice" helped them reduce costs and risks in the production process, which led them to be the first in their industry. World class manufacturing has 10 pillars and each pillar has its own method to identify the best solution regarding improvement. Each problem is treated as a project to demonstrate the results and how much money would they save. They have also revised the company to raise their WCM status to silver.

The production facility consists of a single production line with high level of automation and less human interference in the production process compared to the other three cases.

A world class manufacturing champion was the interviewee in this case. Her job is integrated with different areas in the company's organization because WCM pillars are about 10 different areas including safety, cost deployment, focused improvement, autonomous maintenance, workplace organization, quality control, logistics and customer service, early equipment management, employee development and environment. As a result, she answered different questions regarding the authors' research objectives. As mentioned before, the interviews were held in a semi-structured way to produce new topics and questions about Obeya and the meeting space. The interview took around two hours including a morning meeting and line visitation. Questions asked during the interview were formulated according to research questions, objectives and literature study. Questions were mostly about meeting space characterizations, usage, participant numbers, discussion types, and meeting duration, information visualization and meeting space pros and cons. The interviewees' voice and authors' notes were recorded and documented.

Meeting spaces

A single meeting place is used for morning meetings which were less structured and limited. The meeting place was located near the production manager's office, near the middle of the production line. These meetings could take up to 30 minutes like the one the authors participated in. Around eight people with different specialties, from different departments, participate every day including a maintenance manager, production manager, WCM responsible, logistics manager, technicians and sale manager. However, production problems and solutions, products with defections and causes of product imperfections and Kaizen project follow-ups are the most important topics of daily meetings. In general, everything from participants, position of participants and level of discussion (organizational or operational) is similar to other companies. The only difference is the amount of time dedicated to meetings. It helps engineers and managers to go into detail about their production problems or Kaizen activities. As the authors participated in a morning meeting, all engineers were involved in daily discussion regarding problems, solutions and future plans even though there were not specialists in that area. Swedish is the language spoken among engineers, it is also the official language in the company.

The usual visualization and follow up tools and techniques are used in this meeting space, like A3 reports, follow-up forms and boards. It is nearly the same as in above mentioned companies. A3 reports are the main report tool in lean and TPS, however it can also be the main reporting tool in WCM. Boards and sticky notes were used to write the meeting topics and agenda. Results are attached and written on boards and sticky notes and then moved to suitable paper and repots. In addition to conventional visualization mentioned previously, tool improvement tags are used to mark problem sources in the line. One version of the tag is to mark the most problematic stations or machine. Moreover, a digital screen is used to visualize general information about the company and current situation of production. It should be mentioned that morning meetings were held informally.

Interview details

A WCM Champion was the interviewee in this case. Her job is integrated with different areas in companies organization as WCM areas or pillars are about 10 various areas including safety, cost deployment, focused improvement, autonomous maintenance, workplace organization, quality control, logistics and customer service, early equipment management, people development and environment. As a result, she was able to answer different question regarding the research questions and objectives. As mentioned before, a semi-structured interview was chosen to be flexible and able to come up with new topics and questions during the interview. The interview took more than two hours including the morning meeting and line visitation. Questions asked during the interview were formulated according to research questions, objectives and literature study that had been performed previously. Questions were mostly about characteristics of the meeting space in Gyproc, its usage, participant numbers in morning meetings, discussion types, and meeting duration, information visualization and meeting space pros and cons. The interview was recorded and the authors' notes were documented separately. It should be mentioned that both authors participated in the interview, meeting and production line visit which increased the validation and quality of the research.

Case D, Fuji Autotech

Company description

Fuji Autotech is a medium size company with about 160 employees that manufactures automotive parts for heavy vehicles. It is an integral part of the international group "Fuji Kiko Group." Their commercial vehicle business is known for lean manufacturing and high quality. The company has almost the 60 percent of the European market share of its product.

According to the interviewee, the company was inspired by Japanese quality thinking, whereas working with Failure mode and effects analysis (FMEA analysis), Poka Yoke solutions, physical testing, virtual testing and vision cameras.

The total production volume is 500,000 units per year. The production facility consists of five assembly lines and the international company has several facilities around the world.

Meeting spaces

Each production line produces product for a different customer and consequently, each production line has its own meeting place where the morning meetings are held. Meetings take about 10 minutes and participants are the line operators, safety technician, group leader and quality technician. However, depending on the topic, other people may participate. Meetings begin with controlling the presence of personnel of the respective line. After that any deviations from the previous day's production from the production goal, the remaining goal will be discussed and the first level cause will be followed. There is no following up of problems or documenting them. Then possible safety issues and solutions are discussed and prevention methods are suggested by participants. In this regard, this case is in the very basic level of using of the meeting spaces. Meeting spaces are not used for implementing or following up on improvement projects in production system. They are solely a place for daily control of production volume and deviations from planned volume.

Respectively, visualization tools are also basic. Usual visualization tools are used in meetings including whiteboards, forms and other basic tools. The meetings and visualization tools are only utilized for controlling basic measures like production rate of the previous day and deviations from planned production. Also, safety issues are discussed during the meetings. A table of people present during every shift is on the board and simple charts are used to show weekly and daily production plans and disparities. There is also a safety corner on the board and incident pictures, information and preventing methods are visualized. These issues are the only instances where follow-up is documented through meeting spaces at the factory.

Interview details

At Fuji Autotech, a process development manager was asked to be the interviewee. He had worked for Fuji Autotech more than five years and had good knowledge of and experience with production and its problems in different areas. He was one of the main participants in meetings at Fuji Autotech.

The semi-structured interview was performed like the other cases but with process development manager. Flexibility and discussion during the semi-structured interview were the main reasons to choose this type of interview for the research. The interview at Fuji Autotech took about 1:45 including the interview, production line visit, Obeya meeting and company presentation for almost 10 minutes. However, while the production line and meeting spaces were visited, the process development manager was explaining the structure, goals and application of such meeting places at Fuji Autotech. Questions asked throughout the interview and company visit were chosen according to research questions, objectives and literature study that had been done previously. At each case company, the authors tried to cover new ideas, along with experiences from previous interviews and visits to improve the interview and research quality. Questions asked were mostly about specifications of the meeting spaces, their usage, participant numbers, discussion types, meeting duration, information visualization format and advantages and disadvantages of the meeting space in viewpoint of the process development manager. The whole interview was recorded, just like the other case studies and written notes by the authors are documented. What's more, both authors participated in interview and production line visits to increase the validation and quality of the research.

Table 2- General info about case studies

Company	No. of departments/lines	No. employee	Industry
A	5 assembly lines	1,800	Material handling equipment
B	3 main and 1 maintenance departments	1,300	Automotive
C	Single production line	360 in Scandinavia	Construction material
D	5 assembly lines	160	Automotive parts

Cross-case analysis

Table 3 shows a summary of gathered data about all four cases

Table 3 - summery of gathered data

Company	Meeting place type	Purpose	Visualization tools	Meeting time (Minutes)
A	Single place for all lines	Kaizen projects	Predefined A4 forms, A3 reports, boards	5 – 10
B	Multiple customized places for each department	Kaizen projects, General development projects	Predefined A4 reports in the lines, customized reports, schedules and charts for maintenance department, boards	Up to 10
C	Single room for single production line	Kaizen projects	Predefined A4 reports, problem reporting tags, digital screen for production status, boards	Up to 30
D	Multiple places for each assembly line	Production, quality and safety control	Simple quality and production A4 reports, boards	Up to 10

Meeting space use

As the summary of the results indicates, the main use of meeting spaces in first three cases is managing and following up incremental improvement projects in production system. These types of projects are usually initiated by product defects or process errors. Such defects can be reported by internal and external customers or the line operators themselves. Comparison between the four cases reveals the huge opportunity to benefit from the meeting spaces.

In company D, as the less advanced case, the meeting spaces are only used to control the production and safety issues and improvement is not followed up directly in the meeting space. It is only used to send signals about deviation from production goals to management and production planners. In this regard, it seems the company can benefit from such meeting spaces by using them at least for incremental projects.

In contrast to the maintenance department of Case B some activities were observed which can be interpreted as Kaikaku-supporting activities. Specific visualization tools are used for supporting daily meetings implementing a new maintenance system in the factory. The new maintenance system was part of implementing a lean production system in the company. In this regard, the project can be considered a radical change due to the systematic change in a relatively short time.

In general, the results show such meeting spaces have a vital role in continuous improving production facilities and sharing data about the process, results and lessons learned.

Number of meeting spaces

In regards to the number of spaces, no case uses same system. This could be due to differences in products, production system, number of lines, number of departments and organizational differences.

Company A uses a single meeting place for all five assembly lines, while companies B and D use different spaces for each line or department. Although using different meeting spaces can make each department able to customize its space according to its specific requirements, company B and D did not use this benefit except in the maintenance department in case B.

In general, the following benefits have been considered for a single meeting space:

1. Easier information sharing of problems, solutions and the problem solving process
2. Higher level of cross-functionality in the meetings
3. Reduced time of problem solving partially due to better information sharing

4. Using other departments' experiences in developing visualization and documenting tools
5. Easier access to internal customers or other internal-related actors.

But there are also disadvantages with using a single space including a reduced degree of customization and detail in visualization which turn can lead to reducing problem solving creativity in some cases.

The maintenance department in case B is a good example of such customization, which seems necessary in such cases. Especially when it comes to large and radical improvements such as the maintenance system development in case B, where a more detailed visualization and follow up seems inevitable.

Visualization tools

The following visualization tools are found as common tools representing the improvement processes and results in the cases production systems:

1. A3 reports: In the first three cases A3 reports are used for documentation and follow-up the problems. They are also used as visualization tools and stay on the boards during and after the problem solving or improvement process. The most complete version is used in case A with seven sections in most cases. The problem background, current condition, goal, root cause analysis, countermeasures, effect confirmation and follow-ups. Following these steps, the involved people map the problem and its background, try to find the cause of the problem using methods like five whys and fishbone diagram, and, finally, suggest a solution for it. After implementing the solution, the effectiveness of it will be controlled and follow up actions implemented assuring a proper solution is developed and implemented. The process of using it conforms with the suggested method in literature (Liker, 2003).

2. Whiteboards: They are an inseparable part of visualization methods in meeting spaces because of their flexibility, ease of use and cheap price. They are used for various functions such as listing ideas in brainstorming sessions, visualizing parts and machines sketches, message boards, etc. Whiteboards are basically a place to visualize temporary but useful information. In some cases they also fill the gap of not using digital tools to show changing or temporary information and, in some cases, they are more successful than digital screens because the higher level of interaction with users, simplicity and cheaper price.

3. Problem reporting tags: These tags mark the source of the problem. Every tag has two sheets of 18" x 12" of paper summarizing the problem and its cause. One copy is attached to the board in the meeting room and the other to the

station or machine which is the source of the problem. In this way, the stations or machines causing the most problems are easily marked as "Christmas trees" with lots of tags and the responsible people can easily find which stations or machines require improvement.

This tool is only used in company C and a factor which makes the company able to use this tool is a high level of automation in the line. Highly automated lines and few operators help the company use the problem reporting tags without increasing tension in the work environment. In this regard, using this tool in other companies needs more consideration and staff training to prevent possible tensions.

4. Digital screens: Digital screens are used to show frequently changing data. In case A and C, such screens are used to show data like daily production rate, line stoppage time and elapsed time since the last safety incident in the factory. No further use of any digital tools were observed in the case studies.

5. Customized tools: There are a number of customized tools, especially in the maintenance department in case B, which are used for following-up on larger improvement projects. As mentioned before, this case was the only one with improvement projects that can be considered radical. Some instances of such visualization tools are described below.

Simplified Gantt charts: a wall was assigned to represent a large scale, simplified Gantt chart with project highlights on important delivery times and stage gates, main activities behind the schedule and some other critical information.

Resource allocation chart: there was a resource allocation chart on the other wall to visualize the status of resources, especially human ones, highlighting resource bottlenecks and other resource related status.

For such tools, using more advanced and interactive digital tools seems helpful and supportive. Digital visualization tools can support easy on demand access to different level of details, managing multiple projects in a single space and better information flow. Such issues are discussed more in recommendations section.

Meetings' time

Finally, the meeting time was another factor considered in this study. In all cases, except case C, the time of the meeting were limited to 10 minutes and they started in the morning/before the beginning of the shift. The limited time helped keep the meetings brief, simple and goal oriented. Also, it prevented wasting the time of non-related people. Details of problem or solutions are discussed later, after the meetings and only with related people. But the necessary details are documented in forms and reports which are visualized in the meeting space for other peoples' access.

This way, all people in the meeting have the problems general and access to the solutions through documents and people. They do not need to know all the details, but they have access to them any time, to use as learned lessons and benchmarks in similar cases and problems.

In company C, there was no specific limit for time of the meetings. This could be partially due to having a single production line with a high automation level. Having only one highly automated production line allowed those in the meetings to focus only on that line and there are no non-related people t comparing with the other three cases. However, of the authors' participation experience in the meetings indicate that long meetings can lead to decreased efficiency and deviation from the goal with having peripheral and, in some occasions, unrelated discussions and limited time can be a positive factor limiting the discussions to the most important ones and lead the participants to reach an agreement as soon as they can regardless of unnecessary details.

Recommendations

As mentioned in the analysis section use, using Obeya or similar spaces for developing and supporting radical change in companies could be an opportunity for companies to facilitate and speed up projects.

Having common meeting spaces for departments facilitates the knowledge and experience flow in the companies leading to faster accumulating results and experiences and making them more accessible. However, having customized and separate spaces for specific departments, in some cases, is inevitable.

Limiting meeting time to better concentration on the topic with related people is another point which can be recommended according to the results and analysis.

The last, but not the least, point is considering use of digital tools to facilitate the data transfer and knowledge sharing. Using such tools can increase the efficiency and make it possible to use a single meeting space for different purposes. Although some interviewees pointed out the advantages of handwritten documents and notes, others have the same idea as the authors that the combination of digital tools with traditional tools can boost the meeting space's effectiveness. This issue will be discussed more in the next section on discussion and conclusion.

Chapter 5: Discussion and conclusion

At first, this chapter presents a discussion around the research. Discussion is started by comparing the results of the authors' literatures review and the four case studies. Afterwards, conclusion of the research and answering the formulated research's questions are presented. Finally, possible future work is proposed.

As mentioned before, the research was performed in closed loop between theoretical framework and real life observation. This loop is called theory matching and it seeks to validate the information found in literature by investigating among real life case studies. According to the research questions and objectives of this thesis, four main subjects from literature were chosen to cover different facts about Obeya. These chosen facts are Obeya's type, usage of such meeting spaces at manufacturing companies, applied visualization tools inside the Obeya and meeting's duration. Afterward, the facts found in literature were compared with real status and usage of meeting places or Obeya at manufacturing companies in Sweden. The comparison demonstrates what has been described in literature and what is being done have approximately the same direction expected from two elements that would be described later. On one side, it shows the case studies verify the results of literature and in other side, it validates the quality of the conducted research.

The four main chosen facts about Obeya are shown briefly in Table 4. These facts were chosen to answer the research questions and fulfill the research objectives. However, these four areas can be divided further into more details to get more precise results.

Types of Obeya

The first fact was chosen according to the first thesis objective and the first research question. The authors formulated the practice of meeting spaces in term of number of projects and problems discussed in such meeting spaces. The majority of literature discusses a single room that a group of experts from various departments get together in frequently to review a project and discuss key decisions about that project. However, few authors has written about the usage of Obeya or meeting places but for projects being run simultaneously (Söderberg and Alfredson, 2009; Andersson and Bellgran, 2009). Hence, two different types of Obeya or meeting spaces were defined as follows:

- Single for whole company: as mentioned the primary definition of Obeya is a single room for a single project. A perfect example is the Prius project at Toyota that only used Obeya to develop the Prius and introduce it to market with a shorter lead time.

- Multiple for each department or production lines: For example, Andersson and Bellgran (2009) describes a meeting place in one of her cases as:

"The Project studio was created in 2006. The project studio was originally created in order to let different development projects sit closely to each other and to the production. The aim was to improve the integration between construction and production in order to develop better and more production-friendly products."

Additionally, Söderberg and Alfredson (2009) describe a meeting place in Scania as:

"Another Lean tool that has been implemented is visual planning; Scania has a room called "the pulse room" where the status of all the projects currently running in the company is shown. This is a very powerful tool, since Scania need a way to easily show which projects are going according to schedule and where effort needs to be put. It also shortens meetings substantially and a meeting dealing with more than 100 projects can be over in half an hour. One advantage with having this visual planning is that there is a clear ownership of the projects."

Purpose of using Obeya or meeting rooms

The second fact was selected to answer the second and third questions and fulfill the second thesis objective. As a result, according to performed literature study four features were selected for this fact. These features are

- *Incremental improvement or Kaizen* (Andersson Schaeffer, 2011; Oppenheim, 2004): It refers to use Obeya to support daily Kaizen activities and eliminating the eight wastes in production or Muda. Kaizen concept is well recognized by most companies around the world and various tools and techniques are being developed and utilized to support Kaizen (Liker, 2003; Yamamoto, 2010). Nevertheless, despite of a well conceptualize definition of Kaizen in literature; practice of Kaizen is not well performed within manufacturing industries out of Japan (Yamamoto, 2010). Therefore, Obeya as a lean tool in manufacturing industries can accelerate implementation of Kaizen in industries. Within the Obeya room, all other lean tools can be questioned if they are implemented well enough or not. Obeya would shorten the decision making time between problem recognition and taking action since most involved people are there. Combining Obeya and Kaizen can be referred to Kaizen corner, which was built for education, short daily meetings with managers and group leaders and engaging labors through training and morning meetings (Liker, 2003). According to conducted case studies, 3 of 4 companies use their meeting rooms for solving daily production problems and Kaizen activities. For example, Obeya as Kazien corner in maintenance department at Volvo Powertrain in Köping.

- *Radical improvement or Kaikaku* (Andersson Schaeffer, 2011): It refers to using Obeya or similar meeting spaces for handling radical changes within the company. As explained later, Obeya cannot be placed in radically innovative Kaikaku. On the other hand, Obeya can be placed in infrastructural. As a result, Obeya fits into Kaikaku type II, structural change. Fortunately, Kaikaku type II is most common Kaikaku among companies which have experienced it. Thus, respecting the trend of lean transformation in companies around the world, especially in Sweden, the need of Kaikaku, and consequently Obeya, cannot be neglected. None of the case study uses their meeting places for such purpose. This can be considered a big question for further research.
- *Product development* (Liker, 2003; Andersson Schaeffer, 2011; Lindlof and Soderberg, 2011): Since Prius development, Obeya has become a standard support tool for product development process in Toyota. In the previous vehicle development projects, the chief engineer needed to go around to people's offices in different departments to conform to others. Conversely, in Obeya, cross functional team of experts from production, purchasing and product development gather in a big room to review the progress and discuss key decisions. The aim of such room is to integrate product development and production to have better and more production-friendly products. Case study companies were not using their meeting spaces for such objective. However Volvo Powertrain in Köping use its meeting place for general development projects in a very low level.
- *Others (for example: Safety control):* This option was added for probable use of Obeya or similar meeting spaces that was not mentioned in literature. In one the case studies, meeting spaces are being used for putting and showing safety issues and controlling whether all workers follow it or not.

Visualization tools inside the Obeya

Visualization tools were chosen because of the second objective and fourth research question. This part consists of various tools within continuous improvement like boards and charts as well as tools within project management such as WBS, Gantt charts and network diagrams. However, the below list of tools are the most important ones that were mentioned in nearly all of the literature or tools that are necessary and common in meeting spaces like video projection.

- A4 reports (normal)(Liker, 2003; Lindlof and Soderberg, 2011):

- A3 reports (Root cause and 5 why's)(Sobek and Smalley, 2008; Söderberg and Alfredson, 2009; Liker, 2003):
- Boards (Andersson Schaeffer, 2011; Liker, 2003; Lindlof and Soderberg, 2011; Parry and Turner, 2006)
- Charts and schedules (Andersson Schaeffer, 2011; Liker, 2003; Lindlof and Soderberg, 2011; Parry and Turner, 2006)
- Digital screens (Liker, 2003; Parry and Turner, 2006; Oppenheim, 2004; Lindlof and Soderberg, 2011)
- Video projection (Parry and Turner, 2006)
- Typical things (Lindlof and Soderberg; 2011, Liker, 2003)
- Problem report tags

Duration of meetings

Eventually, the last fact was chosen to answer question four and fulfill the second objective. Meeting duration was not studied and considered in previous literature. It was generally mentioned that Obeya is an advanced visual control innovation room where cross functional groups of expert from various divisions gather on a "*regular basis*" to review progress and discuss key decisions (Liker, 2003; Andersson Schaeffer, 2011; Söderberg and Alfredson, 2009). Nevertheless, in few articles was it mentioned that meetings have to be held at least every other day to get the suitable result.

All case studies in this thesis use their meeting spaces every day. In three cases, the duration of each meeting was around 10 minutes and in one case it was up to 30 minutes. In accordance with literature, the vital part of Obeya is the frequency of meetings which was followed by all case studies.

As the results indicate, in all cases the meeting spaces were used only for performing incremental changes. These modifications are mostly initiated by minor problem occurrences in the production process, product defections or safety issues. Solutions are developed using lean tools like 5whys and 5 Ws for preparing A3 reports. Currently there is no indication for using such meeting spaces in radical changes in production system development, such as developing and implementing new production system or general modification in current production systems. In such cases, production system development is mostly considered a part of the product development process according to its dependence to developing new products(Bruch, 2012). But in practice, it is a huge complex separate project. Obeya meeting spaces can be used for acquiring and generating production system development information like idea development sessions and designing production system. It can also be a very useful tool for sharing and using information especially during the implementation of radical changes in production systems.

Table 4 - comparison between literature and case studies

Literature study	Case studies
Obeya type	
Single for whole company	Single for whole company (Gyproc and BT)
Multiple for each department or production lines	Multiple for each department (Volvo: production and maintenance) or production line (Fuji: 5 line)
Purpose of using Obeya or meeting room	
Incremental improvement or Kaizen	Incremental improvement or Kaizen (BT, Volvo, Gyproc)
Radical improvement or Kaikaku	Radical improvement or Kaikaku (None of them)
Product development	Product development (None of them but Volvo for general development)
Others (for example: Safety control)	Others (for example: Safery control) (Fuji Autotech)
Visualization tools inside the Obeya	
A4 reports (normal)	A4 reports (normal) (All)
A3 reports (Root cause and 5 why's)	A3 reports (Root cause and 5 why's) (BT, Volvo)
Boards	Boards (All)
Charts and schedules	Charts and schedules (Volvo, BT)
Digital screens	Digital screens (Gyproc)
Video projection	Video projection (None)
Typical things	Sticky notes, magnet pins (Gyproc, BT, Volvo)
Problem report tags	Problem report tags (Gyproc)
Meeting time	
Regularly every other day or every day	every day, 10 min (Volvo, BT, Fuji Autotech) every day, 30 min (Gyrproc)

In addition, current meeting spaces are not adequately capable of transferring data and results to involved internal actors like people in other production sites or external ones like suppliers. This could be due to dependence of those spaces on non-digital tools.

As Bruch (2012) explains, designing information management as a critical part of production system design and development consist of three main parts: acquiring, sharing and using information. Obeya can be used for these purposes in the production system design and development process. Using digital tools can help the two latter parts facilitate sharing and using acquired data in a faster and more effective manner.

In summary, reviewing the Obeya concept, its advantages and its current practice in industry shows that it can be applied to purposes other than product development projects. The case studies show that similar meeting places are already used for incremental production system development projects. Radical improvements can even benefit more from this concept because of their nature that needs to implement great changes in a short time usually demanding considerable amount of close teamwork. But to maximize the benefits, methods and visualization tools used a conventional Obeya should be customized to be adapted to this purpose.

Fulfillment of the research objective

In this study, both research objectives were covered though the methodology based on current literature and empirical data. Comprehensive study of current literature has been done about the current uses of Obeya and similar meeting spaces. Also, four case studies have been complete, gathering empirical data and comparing them to results from literature. The empirical data supports literatures results that such meeting places are utilized for daily follow up of incremental improvements in production systems. Such improvements are almost always Kaizen projects, which start and need to be followed up on a daily basis by line operators, supervisors and other related people. In one case, partial use of such meeting spaces for radical change projects were also observed. But using these meeting spaces for Kaikaku projects in production system developments has not been generalized yet.

In addition, some suggestions for improving the meeting space use are provided according to gathered data and cross case analysis in this study. The suggestions are described in recommendations section including limiting meeting time, centralizing meetings in a single space and combining traditional visualization tools with digital ones to facilitate data sharing. Also, broad opportunity for further research in this field has been identified which is described in the next section.

Future work

This research is only an opening to study the supporting role of Obeya in production system development. A few directions for further work in this field are identified by the authors, including:

- Deeper study of meeting spaces role in Kaikaku projects in production system development and opportunities that such spaces can provide in radical development projects.
- The influence, both positive and negative, of using digital visualization tools, on meeting efficiency and facilitating data sharing and use. The degree of digitizing visualization tools can lead to optimum combination of traditional and digital tools to support meetings could be a specific focus in this topic.
- The role of the meeting spaces as a common forum for product and production system developers to support easier and more effective flow of the required information in both processes.
- This thesis has been based on four case studies and their regulation on holding meetings. However, to draw generalizations from this research, the sample needs to be much larger. Perhaps more cases should be considered, whether in Sweden or in other countries.

There are certainly other opportunities to broaden the knowledge of Obeya which can be covered by future research in this field.

Reference

Almström, P. 2005. *Development of manufacturing systems - a methodology based on systems engineering and design theory.*, göteborg, sweden, chalmers university of technology.

Andersson, J. & Bellgran, M. 2009. Spatial design and communication for improved production performance. *Proceedings of the international 3rd swedish production symposium.* Göteborg, sweden.

Andersson Schaeffer, J. 2011. *Communication space: spatial design in manufacturing industry.* Mälardalen university.

Andreewsky, E. & Bourcier, D. 2000. Abduction in language interpretation and law making. *Kybernetes,* vol. 29, pp. 836-845.

Arbnor, I. & Bjerke, B. 1994. *Methodology for creating business knowledge*, sage publications.

Arlbjørn, J. S. & Halldorsson, A. 2002. Logistics knowledge creation: reflections on content, context and processes. *International journal of physical distribution & logistics management,* vol. 32, pp. 22-40.

Bellgran, M. 1998. *Systematic design of assembly systems: preconditions and design process planning.* Phd thesis, linköping university.

Bellgran, M. & Säfsten, K. 2005. Produktionsutveckling, utveckling och drift av produktionssystem. *Isbn 91-44-03360-5, studentlitteratur, lund.*

Bennett, D. 1986. *Production systems design,* london, uk: butterworth-heinemann.

Berger, A. 1997. Continuous improvement and : standardization and organizational designs. *Integrated manufacturing systems,* vol. 8, pp. 110-117.

Bicheno, J. 2004. *The new lean toolbox: towards fast, flexible flow*, picsie books.

Bilalis, N., Scroubelos, G., Antoniadis, A., Emiris, D. & Koulouriotis, D. 2002. Visual factory: basic principles and the 'zoning' approach. *Ijpr,* 40, 3575-3588.

Blanchard, B. S. & Fabrycky, W. J. 1998. *Systems engineering and analysis.*

Bruch, j. 2012. *Management of design information in the production system design process.* Phd phd thesis, mälardalen university, school of innovation, design and engineering.

Brunet, A. P. & New, S. 2003. In japan: an empirical study. *International journal of operations & production management,* vol. 23, pp.1426 - 1446.

Cochran, D. S., Arinez, J. F., Duda, j. W. & Linck, J. 2002. A decomposition approach for manufacturing system design. *Journal of manufacturing systems,* 20, 371-389.

Danermark, B., Ekstrom, M., Jakobsen, L. & Karlsson, J. C. 2001. *Explaining society: an introduction to critical realism in the social sciences* routledge; 1 edition.

Dilokpimol, K. & Surasawadi, n. 2012. *Kaikaku in production system: in case of innovative production development system.* Master, mälardalen university.

Dubois, A. & Gadde, L.-E. 2002. Systematic combining: an abductive approach to case research. *Journal of business research,* 55, 553-560.

Eisenhardt, K. M. 1989. Building theories from case study research. *The academy of management review,* vol. 14, pp. 532-550.

Ellram, L. 1996. The use of the case study study method in logistics research. *Journal of business logistics,* 17.

Ellram, L. M. 1993. Supply-chain management: the industrial organisation perspective. *International journal of physical distribution & logistics management,* 21, 13-22.

Fagerström, B. 2004. *Managing distributed product development. An information and knowledge perspective.* Doctoral, göteborg.

Flint, D. J. & mentzer, J. T. 2000. Logisticians as marketers: their role when customers' desired value changes. *Journal of business logistics,* vol. 21 no. 2, pp. 19-45.

Glaser, B. G. & strauss, A. L. 1967. *The discovery of grounded theory; strategies for qualitative research,* chicago, aldine pub. Co.

Gummesson, E. 2000. *Qualitative methods in management research*, sage.

Hayes, R. H. & wheelwright, S. C. 1979. Link manufacturing process and product life cycles. *Harvard business review,* vol. 57, (133-140).

Imai, M. 1986. *Kaizen (ky'zen), the key to japan's competitive success*, mcgraw-hill.

Jugdev, K., Mathur, G. & Tak, F. Project management assets and project management performance: preliminary findings. Technology management in the energy smart world (picmet), 2011 proceedings of picmet '11:, july 31 2011-aug. 4 2011 2011. 1-7.

Kennedy, M. N. 2003. *Product development for the lean enterprise* virginia, oakela press.

Kovács, G. & Spens, K. M. 2005. Abductive reasoning in logistics research. *International journal of physical distribution & logistics management,* vol. 35 pp.132 - 144.

Kovács, G. & spens, K. M. 2006. A content analysis of research approaches in logistics research. *International journal of physical distribution & logistics management,* 36, 374-390.

Kulak, O., Durmusoglu, M. B. & Tufekci, S. 2005. A complete cellular manufacturing system design methodology based on axiomatic design principles. *Computers & industrial engineering,* 48, 765-787.

Liker, J. 2003. *The toyota way: 14 management principles from the world's greatest manufacturer*, mcgraw-hill; 1 edition.

Lindlof, L. & Soderberg, B. 2011. Pros and cons of lean visual planning: experiences from four product development organisations. *International journal of technology intelligence and planning 2011,* vol. 7, pp. 269-279.

Manufuture. 2006. *Assuring the future of manufacturing in europe* [online]. Available: http://www.manufuture.org/.

Martin, J. E. & Remo, A. B. 2007. Visual representations in knowledge management: framework and cases. *Journal of knowledge management,* 11, 112-122.

Mentzer, J. T. & kahn, K. B. 1995. A framework of logistics research. *Journal of business logistics,* vol. 16 no. 1, pp. 231-50.

Meredith, J. 1998. Building operations management theory through case and field research. *Journal of operations management,* 16, 441-454.

Miltenburg, J. 2005. *Manufacturing strategy: how to formulate and implement a winning plan*, productivity press.

Morgan, J. & Liker, J. K. 2006a. *The toyota product development system: integrating people, process, and technology*, taylor & francis.

Morgan, J. M. & Liker, J. K. 2006b. *The toyota product development system: integrating people, process, and technology*, productivity press.

Näslund, D. 2002. Logistics needs qualitative research – especially action research. *International journal of physical distribution & logistics management,* vol. 32, pp. 321-338.

Olausson, D. & Berggren, C. 2010. Managing uncertain, complex product development in high-tech firms: in search of controlled flexibility. *R&d management,* 40, 383-399.

Oppenheim, B. W. 2004. Lean product development flow. *Systems engineering,* 7, no-no.

Osono, E. 2008. *Extreme toyota : radical contradictions that drive success at the world's best manufacturer / emi osono, norihiko shimizu, hirotaka takeuchi ; with john kyle dorton,* hoboken, n.j. :, john wiley & sons.

Parry, G. C. & Turner, C. E. 2006. Application of lean visual process management tools. *Production planning & control,* 17, 77-86.

Porter, M. E. 1990. *The competitive advantage of nations: with a new introduction*, free press.

Salloum, M. 2010. *Towards dynamic performance measurement systems.* Mälardalen.

Seliger, G., Viehweger, B., Wieneke-toutouai, B. & Kommana, s., r. Knowledge-based simulation of flexible manufacturing systems. The second european simulation multiconference, 1987 vienna, austria pp. 65-68.

Sobek, D. K. & Smalley, A. 2008. *Understanding a3 thinking: a critical component of toyota's pdca management system*, taylor & francis.

Stake, R. E. 1995. *The art of case study research*, sage publications.

Suh, N. P. 1990. *The principles of design,* usa, oxford university press.

Säfsten, K. 2002. *Evaluation of assembly systems: an exploratory study of evaluation situations,* linköping, sweden, division of production systems, department of mechanical engineering.

Söderberg, B. & Alfredson, L. 2009. *Building on knowledge, an analysis of knowledge transfer in product development.* Chalmers university of technology.

Taxén, L. & Lilliesköld, J. 2008. Images as action instruments in complex projects. *International journal of project management,* 26, 527-536.

Taylor, Steven S., Fisher, D. & Dufresne, R. L. 2002. The aesthetics of management storytelling. *Management learning,* 33, 313-330.

Ulrich, K. & Eppinger, S. 2003. *Product design and development*, mcgraw-hill/irwin.

Warnecke, H. J. & Hüser, M. 1995. Lean production. *International journal of production economics,* 41, 37-43.

Westbrook, R. 1995. Action research: a new paradigm for research in production and operations management. *International journal of operations & production management,* 15, 6-20.

Wiktorsson, M. 2000. Performance assessment of assembly systems. *Kth,* institutionen för produktionssystem.

Womack, J. & Jones, D. 1991. *The machine that changed the world : the story of lean production*, new york: simon & schuster inc.

Womack, J. P. & Jones, D. T. 1996. *Lean thinking: banish waste and create wealth in your corporation,* new york, simon & schuster, ny.

Wu, B. 1994. *Manufacturing systems design and analysis: context and techniques,* london, uk: chapman & hall.

Wu, B. 2001. A unified framework of manufacturing systems design. *Industrial management & data systems,* 101, 446-469.

Yamamoto, Y. 2010. *Kaikaku in production.* Phd, mälardalen universuty.

Yamamoto, Y. & Bellgran, M. 2009 kaizen and kaikaku for competitive production. *Mälardalen university, school of innovation, desing and engineering.*

Yin, R. K. 1994. *Case study research- design and methods,* california usa., sage publications inc.

Yin, R. K. 2003. *Case study research. Design and methods,* thousand oaks, ca, sage publications.

Printed by Books on Demand GmbH, Norderstedt / Germany